Underground Cable Fault Location

Underground Cable Fault Location

Barry Clegg

McGRAW-HILL BOOK COMPANY
London · New York · St Louis · San Francisco · Auckland · Bogotá
Caracas · Lisbon · Madrid · Mexico · Milan · Montreal · New Delhi
Panama · Paris · San Juan · São Paulo · Singapore · Sydney
Tokyo · Toronto

Published by
McGRAW-HILL Book Company Europe
Shoppenhangers Road, Maidenhead, Berkshire, SL6 2QL, England
Telephone 0628 23432
Fax 0628 770224

British Library Cataloguing in Publication Data
Clegg, Barry
   Underground Cable Fault Location
   I. Title
621.31934

ISBN 0-07-707804-7

Library of Congress Cataloging-in-Publication Data
Clegg, Barry,
   Underground cable fault location / Barry Clegg.
     p. cm.
   Includes bibliographical references and index.
   ISBN 0-07-707804-7
   1. Underground electric lines. 2. Electric cables—Fault location. I. Title
TK3255.C57   1993
621.319′23′0288—dc20   93-6655
                                                                                                         CIP

Copyright © 1993 McGraw-Hill International (UK) Limited. All rights reserved. No part of this publication may be reproduced, stored in a retrieval system, or transmitted, in any form or by any means, electronic, mechanical, photocopying, recording, or otherwise, without the prior permission of McGraw-Hill International (UK) Limited.

1234 CUP 9543

Typeset by Datix International Limited, Bungay, Suffolk
and printed and bound in Great Britain at the University Press, Cambridge

To my wife Jean, without whose love, tolerance, forbearance and support over many years, this book could not have been written.

Contents

Preface	xi
Introduction	xiii
Acknowledgements	xvii
PART ONE GENERAL	1
1 What is a fault?	3
1.1 The contact fault	3
1.2 The ground contact fault	3
1.3 The break	4
1.4 Ingress of moisture	4
1.5 Crimping	5
1.6 Cross-talk	5
1.7 Flashing fault	5
1.8 Transitory	6
1.9 Partial discharge	7
2 The proper approach	9
2.1 Safety	10
2.2 Diagnosis	10
2.3 Pre-location	12
2.4 Pin-pointing	15
2.5 Confirmation	16
2.6 Recording	16
2.7 Reporting	16

viii CONTENTS

3 Diagnosis and pre-location — 17
3.1 Diagnosis — 17
3.2 Pre-location — 23
3.3 Resistance bridge — 23
3.4 Pulse echo (also called time domain reflectometry) — 42
3.5 Transient methods — 68
Appendix
3A Voltage and current waves in cables — 106

4 Pin-pointing — 112
4.1 Acoustic — 113
4.2 Electromagnetic — 125
4.3 Pool of potential — 128
4.4 Magnetic field — 136
4.5 Audio frequency — 139
4.6 Core-to-sheath faults — 148

5 Confirmation — 152
5.1 Excavation, backfilling and reinstatement — 152
5.2 Exposure — 154
5.3 Analysis of cause — 155
5.4 Recording — 155
5.5 Further testing — 155
5.6 Repairs — 156
5.7 Reporting — 157
5.8 Notes on Part One — 160

PART TWO   RELATED SKILLS AND PROCEDURES — 167

6 Route tracing — 169
6.1 General — 169
6.2 Detection of magnetic fields — 169
6.3 Signal sources — 170
6.4 Methods of connecting/coupling — 172
6.5 Practical approach — 176
6.6 Receiver aspects—tracing, marking and depth checking — 177
6.7 Transmitter aspects – hints and sound practice — 182
6.8 Ground survey — 185
6.9 Metal detection — 188
6.10 Joint location — 190

7 Cable identification — 192
7.1 General — 192

7.2 Methods	194
7.3 Identifying live cables	198

8 Electrical safety — 201
    8.1 Introduction — 201
    8.2 General — 201
    8.3 Safety regulations — 202

PART THREE   SPECIALIZED AREAS — 207

9 Power cables — 209
    9.1 Faults on low voltage systems — 209
    9.2 Faults on medium and high voltage systems — 230
    Appendix
    9A Examples of actual fault locations — 239

10 Telecommunications systems, information and control systems — 245
    10.1 Telecommunications systems — 245
    10.2 Control and information systems — 267

11 Under-floor and under-road, pipe and soil heating systems — 273
    11.1 Types of system — 273
    11.2 Choice of method — 274
    11.3 Practical approach — 276

12 Lighting cables—motorway/highway, road and airfield — 280
    12.1 Fault location philosophy — 280
    12.2 Types of system — 281
    12.3 Methods of fault location — 283

13 Optical fibre cable systems — 288
    13.1 General — 288
    13.2 Types of fibre — 288
    13.3 Specifications and usage — 290
    13.4 Losses, connections and splices — 291
    13.5 Instruments for testing and fault location — 291

PART FOUR   CABLES — 295

14 Information on cables — 297
    14.1 General — 297
    14.2 General construction — 297
    14.3 Voltage withstand of cables — 302
    14.4 Useful tables — 305

PART FIVE   CHOICE OF EQUIPMENT 319
15 Factors 321

16 Choice 325
   16.1 Minimum kit 325
   16.2 Portable equipment versus test trailers and test vans 329
   16.3 Test van specifications—general 331

PART SIX   THE FUTURE OF FAULT LOCATION 335
17 Pressure for solutions, technology and known trends 337

18 Conclusion 343

   Further reading 346

   Index 347

Preface

It has often been said that cable fault location is more of an art than a science. There is a lot of truth in this statement because, although it is a technical subject, all faults are different and success depends to a great extent on practical aspects and the experience of the operator.

Though much has been written on the subject, most of it is in the form of articles and papers presented in the technical press by experts, usually associated with test equipment manufacturers, all of whom quite naturally favour their own approach and equipment. Few books exist devoted solely to cable fault location that cover all known methods.

Engineers engaged in fault location are often handicapped in that:

1 They may not have received thorough training.
2 The range of equipment at their disposal will usually be limited.
3 Detailed information on methods and instruments may well not be available.

The purpose of this book is therefore to pull together as much as possible of the available lore, information and knowledge concerning all aspects of fault location in order to make up a volume of reference which will hopefully be useful for students, engineering managers and field engineers alike. Every attempt has been made to incorporate all methods and approaches, utilizing both historical and modern techniques, together with coverage of related subjects such as cable and pipe tracing, cable identification and safety.*

---

* Neither the author nor the publisher takes any responsibility for the actions of any person in the field. All comments on operations and safety herein are simply guidelines. Operators in the field are bound by the safety rules and regulations of their own organization and country.

Finally, considerable thought has been given to style, layout and content. The graphics and examples are included as *aide-mémoires* and it is hoped that the order of presentation will assist the reader who wants to 'pick' for specific information as well as the one who reads the book straight through.

Introduction

Each fault location is a drama—a detective story in which the sagacious operator makes a careful analysis of the evidence and applies sound logical procedures which lead inexorably and speedily to a correct location . . .

OR . . . each fault location is an embarrassing nightmare of trial and error and miscalculation leading to an *eventual* location.

Interestingly, providing the cable is not abandoned, it is a fact that *all fault*

locations are successful . . . but at what cost in time and money . . . and how many joints are added to the network in the process?

This drama is much influenced by the setting. For example, the cable can be shallow or deep. The route may be over open country, in pavement or roadway, under buildings or rubbish tips, etc. Also, the characters involved have a profound effect on the chances of success. Let us assume that you, the reader, are the fault location engineer. *You are in charge!*

Working with you, alongside you (or, you may sometimes think, against you), are firstly the *management*.

Your boss provides the *resources* such as instruments, transport and the workforce.

Then there is the supervisor who directs operations for you. You have more knowledge than the supervisor but probably less common sense and experience. You make a good team.

Then there are people who dig holes. They have much experience of various types of cable and their habits. They do not like digging in the wrong place.

The *villain* of the piece is the *fault* – smelly, carbonaceous and well versed in the art of avoiding detection!

It is clear that, if you are to find faults quickly and efficiently, you must not only have a good knowledge of the required techniques and methods, but also a thorough grasp of the many other factors involved, such as the allocation of resources, people-management, knowledge of cables, routes and terrain, etc.

Let us hope that this book will help you in this.

Acknowledgements

The author particularly wants to acknowledge the cooperation and counsel of his many friends, colleagues and mentors in many countries around the world. An attempt was made to list them, their locations and affiliations but it was impossible to do this justice. Suffice it to say that they themselves know that they figure in this list and the role they have played. They have the author's sincere thanks and best wishes.

Finally, the author wishes to credit graphics consultant Carole Lucas for her perceptive interpretation of his ideas and Tim Leatherbarrow for realizing them with his brilliant cartoons.

PART ONE

General

1

What is a fault?

A fault could perhaps be defined as any defect, weakness, inconsistency or non-homogeneity that affects the performance of a cable. Although the most common types of fault are either *contact faults* or *breaks*, these can occur in isolation or combination and vary so much in degree that faults can take on a multitude of forms.

The following list is made up of fault types derived from these basic configurations but described also with respect to their anatomy and cause.

1.1 The contact fault

This is also described as a *short* or *shunt fault* and is a connection or part connection between one core and another or others (Fig. 1.1a) or between core(s) and the metallic sheath (Fig. 1.1b).

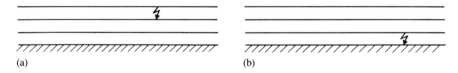

Fig. 1.1 The contact fault (a) core-to-core or (b) core-to-sheath

The value of the fault resistance can, of course, vary between zero ohms and many megohms, a major factor in deciding the location method.

1.2 The ground contact fault

This is variously described as an *earth fault, sheath fault* or *serving fault* and often occurs on unshielded multicore control and telecommunications cables and plastic low voltage cables with no metallic sheath or armour. It is the contact between a core or cores and the mass of earth (see Fig. 1.2a). When

4 UNDERGROUND CABLE FAULT LOCATION

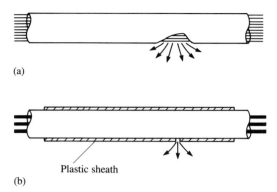

Fig. 1.2 Ground contact fault (a) between core(s) and the mass of earth or (b) between a metallic sheath and the mass of earth

an overall plastic sheath is damaged, ground contact can occur between a metallic sheath and the mass of earth (Fig. 1.2b).

The detection, location and repair of such serving faults is vital in preventing metal sheath corrosion leading to loss of oil (in oil filled cables) and/or core faults with the consequent damage and disruption to supplies or traffic. As a matter of interest, such faults which involve contact between a metal line and the mass of earth lend themselves to almost 100 per cent successful pin-pointing using high voltage d.c. or audio frequency methods. Conversely, pre-location can be very difficult as pulse echo/time domain reflectometry techniques cannot be used.

1.3 The break

This is also called an *open-circuit* or *series fault* and can be a 'clean' break in a conductor, i.e. with an infinite or a very high resistance reading across the break and to adjacent metal (Fig. 1.3a). There can also be a 'dirty' break where there is a measurable resistance across the gap and/or to adjacent metal (Fig. 1.3b). A partial break can occur when some of the strands of a conductor are broken or burnt through (Fig. 1.3c).

1.4 Ingress of moisture

Moisture usually produces a contact fault involving all cores. Water enters a cable at some point of damage and may be present in one limited stretch or it may spread many metres along the cable, typically as far as the next joint. Therefore the site of the breakdown is often some distance from the point of entry, particularly if the cable slopes (see Fig. 1.4). Plastic cables can often perform when wet; paper cables cannot. Insulation resistance values vary considerably but tend to be of the order of a few kilohms. A change of characteristic impedance occurs at the wet section.

WHAT IS A FAULT? 5

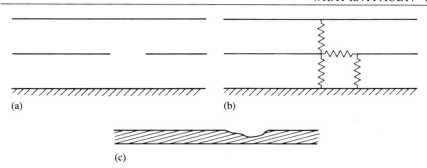

Fig. 1.3 Break (a) 'clean', (b) 'dirty', (c) partial

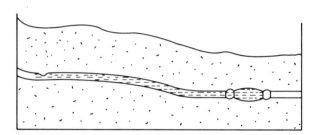

Fig. 1.4 Ingress of moisture

1.5 Crimping
This squashing of a cable is mainly a problem on telecommunications networks as transmission can be affected. In power cables it can simply go unnoticed or, if severe enough, can cause breakdown, thus producing a contact fault. Again, a change of characteristic impedance is noticed which can facilitate location by the pulse echo/time domain reflectometry method.

1.6 Cross-talk
Cross-talk is the picking up on one pair of audio traffic on another pair due to electrical coupling between them. The coupling is normally due to a 'split' or transposition of cores between pairs (Fig.1.5).

1.7 Flashing fault
This is the type of fault that does not manifest itself at lower voltages but flashes over at a certain higher voltage threshold. This may be hundreds of volts or several kilovolts up to a maximum which is the accepted d.c. test voltage for the cable. Such a fault is acting like a spark gap (see Fig.1.6).

6 UNDERGROUND CABLE FAULT LOCATION

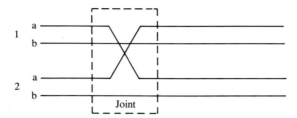

Fig. 1.5 Cross-talk

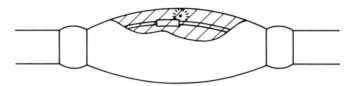

Fig. 1.6 Flashing fault

1.8 Transitory

This is often called a 'nibbling' or 'pecking' fault and usually occurs on low voltage networks; it causes a fuse to blow and then disappears! In fact such a fault is arcing over intermittently, the severity and frequency of this arcing depending on the applied voltage (near or at voltage peaks) and the state of the fault path, i.e. short/long, partly/heavily carbonized, wet/dried out, etc. (see Fig.1.7a and b).

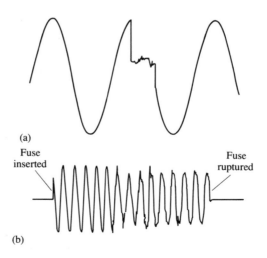

Fig. 1.7 Transitory fault (a) on one half cycle, (b) becoming persistent and causing rupture

It is not often realized that many such faults are constantly active on low voltage cables in most urban areas, but do not necessarily rupture fuses. The arcing only occurs on occasional cycles or groups of cycles until such time, after several cycles/seconds/minutes/hours/days or months, as the fault current has been flowing for long enough to rupture the fuse. It should be borne in mind also that, on a distributor fused at, say, 500 A, a total (fault plus load) current of approximately 650 A must flow for a few minutes before the fuse blows. In other words, considerations of fault level or, more particularly, cable length – and therefore impedance to fault – are very relevant. Indeed, on a badly planned system where the cable is too long and of too small a cross-section, it is possible to have a permanent fault near the end slowly burning its way back along the cable!

A 'dead short' is rare. There is invariably resistance in the fault giving rise to arc drops of tens of volts and sometimes over a hundred volts.

1.9 Partial discharge

Partial discharge is a well-known phenomenon occurring on high voltage a.c. systems whereby small breakdowns occur at weak points in the insulation, e.g. voids. These discharges take place singly, in bursts or continuously, and do not cause breakdown. Their energy is measured in picocoulombs

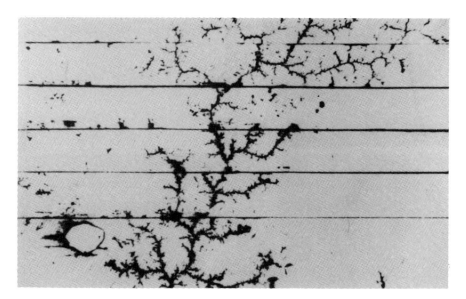

Fig. 1.8 Treeing, i.e. carbon tracking, on paper insulating tapes (*source: Electric Cables Handbook*, Fig. 2.8, p. 23, ed., D. McAllister, Granada, London, 1982)

(pC). Energy levels and general activity invariably increase with time until breakdown does occur. Then there is a fault proper.

'Treeing' is a common manifestation of partial discharge activity. It is produced by the lengthening incursion paths of the arcs and an example is shown in Fig.1.8. It is probable that the future of fault location is bound up in on-line monitoring for the presence and pre-location of partial discharges.

2

The proper approach

It is absolutely vital that a fault location be approached in the correct way, that is to say, *carefully* and *logically*. There is a correct order of doing things which should never normally be deviated from. To do so is to risk:

- Coming to wrong conclusions
- Destroying a fault condition that would have facilitated a location

For instance, there are those in fault location whose first action is to *burn* the fault. They have high voltage and power and they wish to get moving and *do* something.

Faults do not mind being burnt – that is how most of them came into being! Though burn-down is occasionally necessary, it can also destroy a

condition that might otherwise have been very useful in pre-locating or pin-pointing the fault.

2.1 Safety

Electricity is lethal. Few people get more than one chance. Although Chapter 8 covers electrical safety more fully, it is worth considering the following basics as a fault location is about to be commenced:

1. No *action* should ever be taken on a *signal*, pre-arranged or otherwise, or after a pre-arranged *time lapse*.
2. A conductor should never be *touched* until it has been *earthed* ... a conductor should never be *earthed* until it has been *tested* ... and the *tester tested* before and after.
3. All cables and equipment should be *discharged* after testing (even if automatic discharge is incorporated).

The foregoing is a set of absolute basics and is not meant to be a complete safety charter. At all times the safety rules of the authority, organization or company concerned should be strictly adhered to. (Attention is drawn to the safety disclaimer in the Preface of this book.)

The steps in a correct approach are:

- Diagnosis
- Pre-location (treating fault if necessary)
- Pin-pointing
- Confirmation and re-test
- Recording
- Reporting

Sometimes it is impossible, or indeed undesirable, to carry out all the above steps or to keep to this sequence; e.g. pre-location or pin-pointing may alone give a result (or be impossible). Some essential equipment may be unavailable etc.

However, such exceptions are dealt with in detail later. The proper procedure is as follows.

2.2 Diagnosis

Let us again assume that you, the reader, are organizing the fault location. You should bear in mind that if your diagnosis is inadequate or wrong you will fail. Should you succeed despite a poor diagnosis it will only be through luck. It should be your intention to eliminate the element of chance as far as possible, and then be grateful for any lucky break that comes your way!

THE PROPER APPROACH

You should proceed slowly and carefully as follows:

1. *Ask* the people on site about *everything* to do with the situation, e.g. type of cable, length, time in service, information from customers, how the fault occurred, protection operation, any tests carried out already, etc.
2. *Believe* only what is undeniably or most probably true, e.g. type of cable, history, etc., and keep an open mind about other information, for instance the *reported length* of the cable. This is often wrong.
3. *Do not believe* any *opinion* to do with diagnosis or possible location. The responsibility is yours in *success* . . .

. . . or *failure*.

4 *Check* everything yourself.
5 *Test* the insulation resistance between all conductors and sheath and record all values and combinations.
6 *Look* at the cable with a pulse echo set/time domain reflectometer.
7 *Write* everything down. It will be well worth your while to create similar log sheets to the ones shown in Figs 2.1 and 2.2.

This will accomplish two things:

1 During the fault location it will keep your mind clear and assist you in making correct logical decisions and avoiding mistakes.
2 It will provide an invaluable record for the future for you *and others*.

For instance, knowing the type of cable will assist your thought processes. Can an XLPE (cross-linked polyethylene) cable be burnt down? No. Would you expect to find a core-to-core fault on a cable with screened cores? No.

If a cable has been in service and has failed on pressure test there has never been any system fault current and you are almost certainly faced with a high resistance or flashing fault. Knowing the type of protection which operated and whether any auto or manual reclosures have been made will assist you in building up a picture of the fault as your diagnosis progresses.

Would you normally expect a zero ohm fault on one phase of an unearthed high voltage cable system? No. The log sheet will thus provide a detailed record of every scrap of information that can help you, and you need all the help you can get.

You now have enough information to enable you to choose the best methods of pre-location and pin-pointing.

2.3 Pre-location

A pre-location is any test that is carried out from the cable termination and results in a calculation of the distance to the fault, either as a percentage of route length or directly in units of distance. It is sometimes necessary to modify the fault by the application of burn-down voltage in order to create conditions more suitable for a particular pre-location method. However, as a basic rule, it is best to pre-locate the fault with the conditions *as found* if at all possible.

Even though every effort is made to carry out the test with great accuracy, the result will usually be approximate, particularly on a long cable. This is due to inherent errors in the actual test and errors in the measured length of the cable over its route. For instance, a good pre-location may be accurate to within 0.1 per cent, thus enabling the fault to

FAULT LOCATION LOG SHEET, POWER CABLES

PLACE: _____ DATE: _____ FAULT No. ___
CABLE: Type, _____ Length, _____
FROM: _____ TO: _____
SKETCH:

SYSTEM: _____ FAILED: _____
TESTS:

Connection	Continuity	END A		END B	
		R, Ohmmeter	R, Megger	R, Ohmmeter	R, Megger
1-E					
2-E					
3-E					
N-E					
1-2					
2-3					
3-1					
1-N					
2-N					
3-N					

ACTION: _____

PRE-LOCATION DETAILS: _____

PIN-POINTING METHOD: _____

LOCATED AT/REMARKS: _____

Fig. 2.1 Fault location log sheet, power cables

14 UNDERGROUND CABLE FAULT LOCATION

FAULT LOCATION LOG SHEET, TELECOMMUNICATIONS CABLES

PLACE: _____ DATE: _____ FAULT No. ___
CABLE: Type, _____ Length, ___
FROM: _____ TO: _____
SKETCH:

SYSTEM: _____ FAILED: _____
TESTS:

Connection	Continuity	END A		END B	
		R, Ohmmeter	R, Megger	R, Ohmmeter	R, Megger
-E					
-E					
-E					
-E					
-E					
-E					
-E					
-E					
-E					
-E					
Between					
-					
-					
-					
-					
-					
-					
-					
-					
-					

LOCATION METHODS/REMARKS: _____

Fig. 2.2 Fault location log sheet, telecommunications cables

be exposed with one small excavation on a cable route of only a few hundred metres. However,

1. What if there is an unknown loop of cable making the route length measurement inaccurate?
2. What if a section of cable of different size and type has been jointed in and not recorded?
3. What if the cable is 10 km long?

With most methods of pre-location an error of between 0.1 and 0.5 per cent or a few metres can be expected. Therefore, in most cases, a pre-location can only be expected to give a 'suspect area' within which the fault lies.

2.4 Pin-pointing

Pin-pointing is the application of a test that positively confirms the exact position of the fault. By definition such tests are carried out directly over the cable, the most common methods being:

- Shock wave discharge (for power cables)
- Audio frequency (for telecommunications cables and, in certain cases, for power cables)

In the case of shock wave discharge, high voltage (HV) capacitors are repeatedly discharged into the fault. The energy dissipated at the fault position creates a noise which can be heard unaided or detected with a sensitive ground microphone and amplifier, as shown in Fig. 2.3.

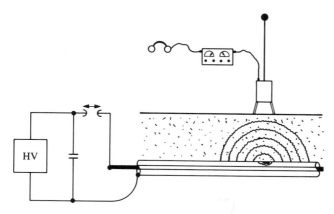

Fig. 2.3 Shock wave discharge – showing sensitive ground microphone and amplifier

Audio frequency methods involve injecting signals of between several hundred and several thousand hertz into the faulty core(s). A search coil and amplifier are then used to detect a change or cessation of signal over the fault. In the case of power cables this can be carried out overground with cables at normal depth, but telecommunications cables usually need to be exposed so that a small pick-up coil can be used directly on the cable.

These and other methods are discussed thoroughly in Chapter 3.

2.5 Confirmation

Of course the fault must be confirmed visually and, in most cases, the site is quite obvious because of external signs such as cracks, breaks, burning and general damage. In some cases there may be no visible evidence, the fault being hidden inside a sound-looking cable. The cable should then be opened up and examined carefully.

After the fault has been 'cleared', the cable should be tested again to make sure that there are no other faults. The existence of more than one fault is rare on power cables but not uncommon on signal, control and telecommunications cables. It is also possible that the cable in the vicinity of the fault is in a poor condition due to moisture travelling each way from the fault site or that tracking has occurred in more than one spot. When confirming a cable fault it always pays to excavate and cut with subsequent repair jointing in mind.

Finally, it is good practice to photograph the faulty cable either for the record or to illustrate bad cable laying, poor jointing practice or, indeed, to provide evidence to claim against third parties who may have caused the fault by damaging the cable.

2.6 Recording

The fault repair, i.e. the joint(s) or any new cable let in, should be carefully noted on the cable records. This is also an opportunity to record the presence of other cables, pipes or obstructions encountered while excavating.

2.7 Reporting

Some authorities do not require the formal reporting of faults while others have very strict requirements which involve the detailed recording of aspects such as:

- Time of occurrence
- Cause/probable cause
- Age and type of cable

The statistics thus gathered can influence the planning, development and maintenance of the system.

3

Diagnosis and pre-location

3.1 Diagnosis

Following the general approach laid out in the previous chapter, the fault location engineer will have found out and recorded all the facts and details relating to the faulty cable and the circumstances of the outage. Figures 3.1 and 3.2 are examples of typical initial notes.

Insulation resistance tests must now be applied to the cable cores to establish the *electrical integrity* of the cable. It is essential that the first tests be carried out with an ohmmeter using a low test voltage of just a few volts. There are two very good reasons for this:

1 Very soon a pulse echo (PE)/time domain reflectometry (TDR) test set will be connected to obtain an *initial 'picture'* of the faulty cable. The application of an insulation resistance test using several hundred or

FAULT LOCATION LOG SHEET, POWER CABLES

PLACE: Hounslow   DATE: 10-10-85   FAULT No. 52
CABLE: Type, 3c Cu PILCSTA, 6-6kV   Length, 3274m
FROM: T.W.A. S/S Jones St   TO: Reservoir Main Switch Room
SKETCH:

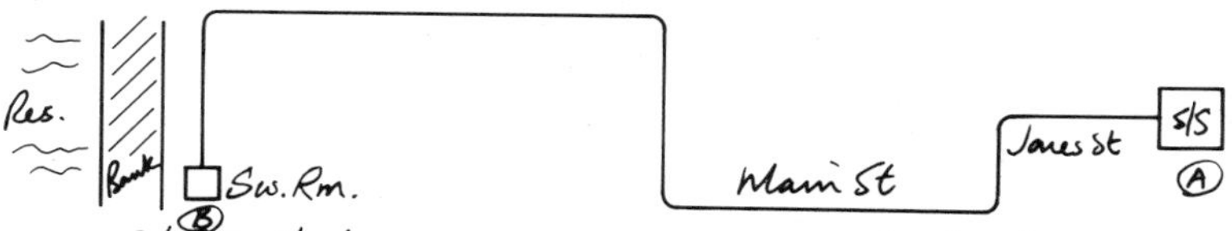

SYSTEM: 3ϕ Earthed   FAILED: In Service

Fig. 3.1 Typical initial notes (power cables)

several thousand volts gives a false impression of the state of the cable *as seen by a PE set/TDR*. That is to say, a medium or even a high resistance fault (which could be a tiny gap) may be shown as *faulty* by the high voltage insulation resistance (IR) tester because its voltage can break down a path, but the PE set/TDR will show no sign of a fault. The result is a confusing and apparently illogical situation.

2 The application of the high voltage from the IR set may modify the fault condition, as the passage of only a few milliamps can sometimes change the state of the fault or make it unstable. Therefore the low voltage ohmmeter must be used initially to test the resistance between all conductors and between each conductor and earth or sheath. If this gives no apparent fault condition, i.e. infinity readings, then the IR tester (or, occasionally, the high voltage (HV) test set) should be used to indicate the faulty core(s).

Figures 3.3 and 3.4 are examples of some typical test results. Sometimes readings differ when the polarity of the test leads is reversed. This is due to electrolytic action at the fault and often indicates the presence of moisture. When these initial resistance tests have been carried out and noted down, it will be apparent which cores are faulty and to what degree, i.e. high/low resistance to earth/between cores, and whether there are any open circuits.

At this point it is not a bad idea to make a sketch of the possible fault situation, as shown in Fig. 3.5a and b. Referring to Fig. 3.5a and b, it should be stressed that they are *possible* configurations. Although faults on different cores are *usually* at the same point, this can by no means be assumed. For instance, in Fig. 3.5a, the shunt faults may or may not be

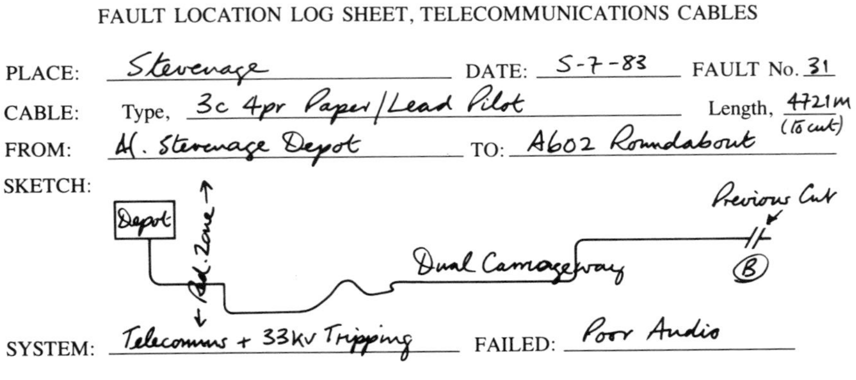

Fig. 3.2 Typical initial notes (telecommunications cables)

DIAGNOSIS AND PRE-LOCATION 19

FAULT LOCATION LOG SHEET, POWER CABLES

PLACE: Watford   DATE: 18-8-84   FAULT No. 40
CABLE: Type, 3c Cu 0.150" PILCSWA, 11KV   Length, 1,892m
FROM: E Rd. Intake s/s   TO: Pumping Station A41
SKETCH:

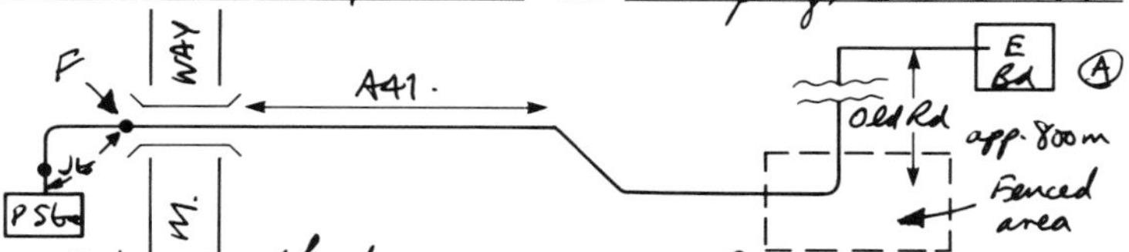

SYSTEM: 3ϕ Earthed   FAILED: Pressure test after jointing
TESTS:

Connection		Continuity	END A		END B	
			R, Ohmmeter	R, Megger	R, Ohmmeter	R, Megger
R	1-E	✓	∞	5M (Vrmg)		
Y	2-E	✓	∞	∞		
B	3-E	✓	∞	∞		
	N-E					
R	1-2 Y	✓	∞	∞		
Y	2-3 B	✓	∞	∞		
B	3-1 R	✓	∞	∞		
	1-N					
	2-N					
	3-N					

ACTION: Continuity checked with Pulse Echo, R ϕ memorised (showing end, no fault

PRE-LOCATION DETAILS: Breakdown intermittent res burnt down to 10Ω - Pulse Echo OK - Location 1775 M
PIN-POINTING METHOD: Shock Wave Discharge - Noisy environment but good result in grass verge past bridge
LOCATED AT/REMARKS: Located at 1770M in new joint.

Fig. 3.3 Example of typical test results (power cables)

at the same point. The open circuit fault is at the same point or beyond it (looking from end A), because a phase 3-E test at end A shows a path and a phase 3-E test at end B shows infinity.

20 UNDERGROUND CABLE FAULT LOCATION

FAULT LOCATION LOG SHEET, TELECOMMUNICATIONS CABLES

PLACE: Bushey   DATE: 17-5-83   FAULT No. 25
CABLE: Type, 15 pr. Cu, 0.63~(10lb/mK) plastic non-filled SW4   Length, 1,362 M
FROM: Trewins Hole (A)   TO: Bushey Hall Fm Hole (B)
SKETCH: Previous cuts made by Water Co

(sketch: 275 KV Line over; water pipe + 2 x 6.6 v.v cables; points A and B)

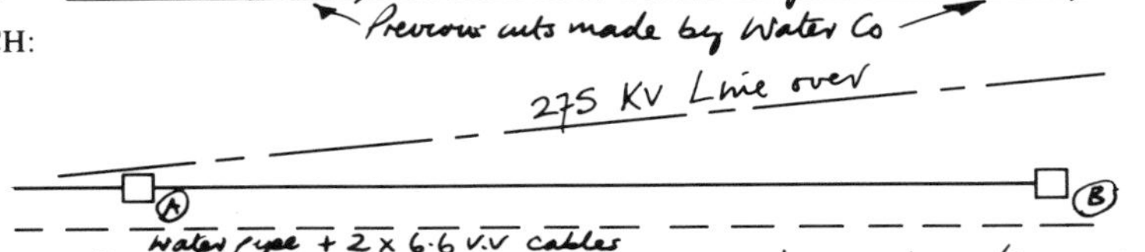

SYSTEM: Control + Telemetry Circuit   FAILED: Low readings (Some cuts made)

TESTS:

Connection	Continuity	END A		END B	
		R, Ohmmeter	R, Megger	R, Ohmmeter	R, Megger
R(RY) -E	✓	20 M		20 M	
Y(RY) -E	✓	18 M		18 M	
B(BW) -E	✓	10 M		10 M	
W(BW) -E	✓	10 M		10 M	
-E	(Many cases with earth rdgs - Lowest 10 M)				
G(BG) -E	✗	10 M		15 M	
Bk(Bk) -E	✗	12 M		15 M	
-E					
-E					
Between Black(BG).G(Bk)	Many readings between legs - lowest 1.6 M				

LOCATION METHODS/REMARKS: High Impedance Bridge used - lowest value 1.6 M
Healthy (orange of Blue/orange)
Faulty (Black of Black/Grey) ⌐ 1.6 MΩ
Return (Grey of Black/Grey) ⌐

between two legs of Black Grey pair... Grey used as return instead of earth. Location at 607.7 M. - Confirmed at 609.5 M
P.E. Set used on o/c leg - gave location of 2nd fault at 952 M - confirmed at 1001 M.

NG

Fig. 3.4 Example of typical test results (telecommunications cables)

Continuity testing is normally carried out by 'shorting' the conductors at the remote end to earth and/or between cores. If a circuit breaker or isolator is closed in the *earth* position, it makes sense to carry out a

DIAGNOSIS AND PRE-LOCATION 21

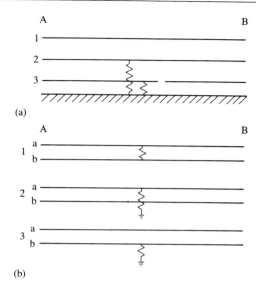

Fig. 3.5 Example sketches of possible fault situations

continuity test before it is opened to fully isolate the cable for further tests.

Again, a low voltage ohmmeter, not an IR tester, should be used and careful attention should be paid to the values read off. These should be in keeping with the loop resistance of the path being tested, correct values for which can be checked from cable statistics such as those given in Part Four, Chapter 14, i.e. fractions of an ohm on power cables and ohms, tens of ohms or hundreds of ohms on telecommunications cables. Any imbalance between readings on different cores or legs indicates the presence of a series fault with some resistance (a 'dirty break').

At this stage of the diagnosis ideas will be forming and it is most useful to *'have a look'* at the cable with a PE set/TDR if one is available. Of course, if a fault is visible, then a pre-location will have been made. However, if there are no open circuits and any contact faults are of too high a resistance, no fault will be visible. The operator will nonetheless begin to *get to know* the cable. The end(s) will be visible and the given length can be checked. Also, significant trace features may be visible, which give an idea of the existence and location of straight joints and tees.

Note that if all cores are known to be continuous and an apparent open circuit trace feature is evident, *the sheath is broken*. There is indeed an open circuit, but not on a core.

Fig. 3.6 'Before' and 'after' traces superimposed

It is also common practice to use a PE set/TDR to check for continuity by watching the end feature on the trace as a short is applied at the remote end. Although this is a useful check, it must be noted that the situation will be unclear if a 'dirty break' is involved.

Notes can be made on the rough sketch (Fig. 3.5a and b) of any possibly significant trace features, e.g. straight joint?, tee joint?, fault? Providing a PE set/TDR with *memory* is available, it is *absolutely essential* that a trace of a chosen faulty core be retained in the instrument's memory.

Hereafter *anything* that the operator does, such as burning down, replacing fuses, surging, etc., may (even fortuitously) change the fault condition. Even a small difference between a subsequent trace and the memorized one 'in the bank' *is the fault*. That is to say, the change in the trace *before* and *after* has been caused by the action(s) mentioned above. This is a very powerful technique and one that should always be employed when the right instrumentation is available. Figure 3.6 shows 'before and after' traces superimposed.

What a shame if no initial trace is taken and the fault condition is

subsequently changed! The operator does not know that the fault could have given itself up there and then and much time could have been saved.

3.2 Pre-location

The diagnosis is now complete and, at this point, the pre-location method can be chosen. This choice depends mainly on:

- The fault conditions
- The instruments available

  and, to some extent, on:

- The cable length
- Accessibility of route and terminations

Difficult or special cases involving factors such as very long or very short cables, lack of equipment, no access to one end, etc., are dealt with in Chapter 9. However, the basic alternatives are:

- Resistance bridge (with or without burn-down)
- Pulse echo/time domain reflectometry (with or without burn-down)
- A 'transient' method such as impulse current or arc reflection

3.3 Resistance bridge

Though considered by many to be old-fashioned, bridge methods can be as effective today as they were many years ago and are often used on medium and high resistance faults when neither PE/TDR nor transient methods can be used. Resistance bridges are applicable to contact (shunt) faults only, the measurement being made by balancing two internal resistance arms against the two external resistance arms represented by the lengths of cable conductor up to the fault.

In order to create the external loop a zero ohm short circuit must be made at the remote end between a healthy core and a faulty one, as shown in Fig. 3.7. This configuration is called the Murray loop test.

At balance, the ratio of the internal resistance arms is the same as the ratio of the external resistance arms:

$$\frac{a}{b} = \frac{x}{y} \quad \text{and} \quad \frac{a}{a+b} = \frac{x}{x+y}$$

and, as the cable has a constant resistance per unit length, x and y can be considered as lengths. The location of the fault as a fraction of *loop* length is given by

$$x = \frac{a(x+y)}{a+b} \quad \text{or} \quad x = \frac{a}{a+b}(2L)$$

24 UNDERGROUND CABLE FAULT LOCATION

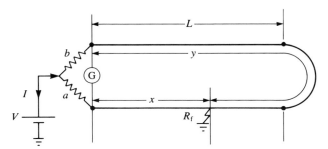

Fig. 3.7 Murray loop test

This holds true if measurements are indeed being made in ohms as with a decade switching system. However, when a potentiometer is used, the dial reading is normally a percentage of the loop length $2L$.

Obviously, as the location of a fault depends on the linearity of resistance per unit length along the cable cores, this brings into focus the main disadvantage of the test, i.e. the cross-sections of the cores, and the positions of any points of change of cross-section must be known. Also, it is clearly necessary to have access to and visit the remote end of the cable in order to loop the healthy and faulty cores.

The main advantage of the Murray loop test is its accuracy. Although any bridge location should be better than 0.5 per cent accurate, it is perfectly possible to achieve accuracies of around 0.1 per cent if there are no other sources of error, e.g. poor route length measurement.

It will be seen that a d.c. voltage is required to pass current through the fault and that, at balance, the galvanometer indicates zero volts.

Two factors influence the sensitivity of the bridge:

- The sensitivity of the galvanometer itself, i.e. what voltage per scale division

and

- The current being driven through the fault

Regarding the first factor, older models have very sensitive moving coil galvanometers with a movement lock for transportation. Also, during measurement, this type of galvanometer is switched into circuit with a key (press button) to save it from being damaged by large deflections when the bridge is far from balance.

Many bridges have an electronic voltage amplifier in place of the galvanometer such as the Megger 18C telecommunications bridge from AVO International, shown in Fig. 3.8. With this bridge, which is used

DIAGNOSIS AND PRE-LOCATION 25

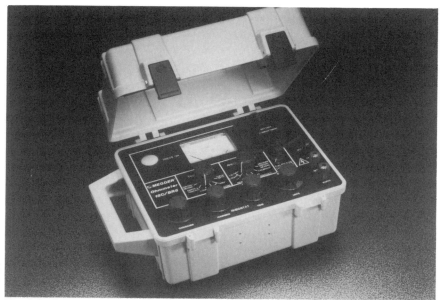

Fig. 3.8 Megger 18C telecommunications bridge (*source:* AVO International, UK)

extensively for telecommunications fault location, variation of the internal resistance arms is made by decade switching. In other bridges this is carried out using a multiturn potentiometer as in the high voltage bridge from Biccotest Limited, depicted in Fig. 3.9.

Referring to the second factor above, the current being driven through the fault is the applied voltage divided by the total resistance of the test circuit, the main component of which is usually the fault resistance. In practice a fine balance is easy to achieve with tens of milliamps flowing. It is more difficult to effect a balance with just a few milliamps and very difficult or impossible with less than one milliamp, at least with the equipment described thus far.

It is evident that better (finer) balances can be achieved by either raising the applied voltage or lowering the current level at which a zero can be detected. Table 3.1 gives applied voltages related to specimen fault resistances and the currents that would flow through them. From this it will be seen that, if a location can be made with currents as low as, say, 0.5 mA, high resistance faults of several megohms can be located with relatively low applied voltages.

Some modern instruments have been developed that operate at these levels. Although connected in the same way and subject to the same

26 UNDERGROUND CABLE FAULT LOCATION

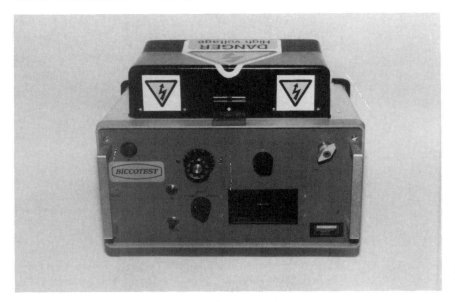

Fig. 3.9 High voltage bridge (*source:* Biccotest Ltd, UK)

Table 3.1 Relationships of applied voltages to fault resistances and currents

Applied voltage (V)	*Fault resistance* (Ω)	*Current through fault* (mA)
2.5	500	5
25	5k	5
250	50k	5
2.5k	500k	5
5k	1M	5
10k	2M	5
0.5	500	1
5	5k	1
50	50k	1
500	500k	1
5k	5M	1
10k	10M	1
2.5	5k	0.5
25	50k	0.5
250	500k	0.5
2.5k	5M	0.5
5k	10M	0.5
10k	20M	0.5

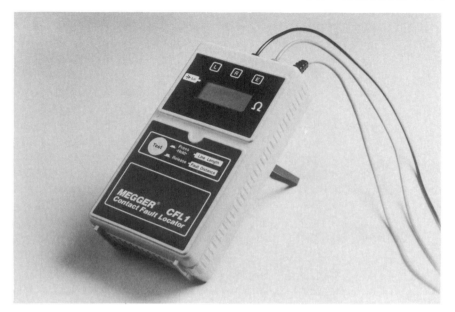

Fig. 3.10 The CFL1 (*source:* AVO International, UK)

constraints as bridges, they are not bridges in that they do not balance. Figure 3.10 shows such an instrument, the CFL1 from AVO International, for use on telecommunications fault locations.

An instrument developed for power and telecommunications cable fault location is shown in Fig. 3.11, the FT1 from Relay Engineering Services Limited.

The anatomies and applications of these recently developed instruments will be set out later. It is first necessary to look at resistance bridge fault location in more detail.

TEST LEADS
Consider Fig. 3.12. If a single pair of test leads is used, this will introduce the resistances of these leads into the external loop as shown. The apparent resistance to fault will then be given by

$$x + l = \frac{a(x + y + 2l)}{a + b}$$

Therefore, if l is very small with respect to x and y, the error will also be small. (Clearly, if the fault lies very close to end A, this error represents a much higher percentage.)

28 UNDERGROUND CABLE FAULT LOCATION

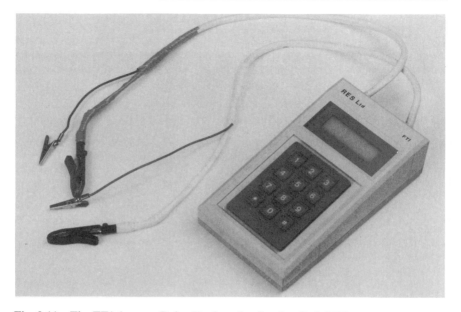

Fig. 3.11 The FT1 (*source:* Relay Engineering Services Ltd, UK)

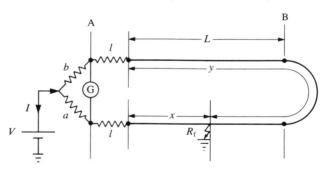

Fig. 3.12 Test leads

This can be the case with telecommunications faults. For example, on a 0.5 mm copper pair 500 m long, the loop resistance would be 87 ohms and if

- a 10 ohms
- b 40 ohms
- x 17.34 ohms
- y 69.66 ohms
- l 0.1 ohm

the apparent resistance to fault is

$$x + l = \frac{10 \times (87 + 0.2)}{50}$$

$$= 17.44 \text{ ohms (or 200.46 m)}$$

This is actually made up of the real resistance to fault 17.34 ohms, plus the resistance of one test lead, 0.1 ohm, which represents an error of 1.15 m in 200,46 m, i.e. approximately 0.6 per cent. Therefore it can be said that test lead resistance does introduce some degree of error into measurements on telecommunications cables, but need not be too important a factor unless extreme accuracy is required.

However, the matter of test lead resistance in power cable fault location is of the utmost importance and must always be accounted for. Say a fault has to be found on a 1500 m long 33 kV cable with copper cores of 185 mm² cross-sectional area and a loop resistance of 0.2937 ohms (using a core resistance of 0.0991 ohms/km). Therefore a single lead resistance of 0.1 ohm represents 1009 m of core – about one third of the loop length of the cable under test! This serves to emphasize that the short at the remote end, B, *must* be a short, heavy, bolted connection of *zero* ohms.

It is possible to use connection leads of large cross-section and account for their equivalent length in the calculations but, in almost all cases in practice, this is not necessary because a 'four-wire test' is used. This simply involves bringing out the galvanometer connections separately to the cable termination, as shown in Fig. 3.13. The main connections are usually flexible conductors of substantial cross-sectional area such that their resistance is negligible compared to that of the internal resistance arms. Therefore an accurate result can be obtained.

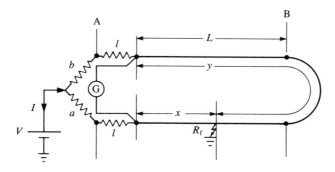

Fig. 3.13 'Four-wire test'

30 UNDERGROUND CABLE FAULT LOCATION

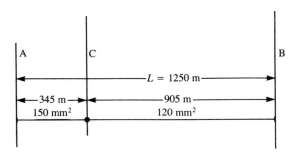

Fig. 3.14 Example cable showing lengths of different cross-section

CROSS-SECTIONAL AREA
When a cable is not of the same cross-sectional area throughout its length, the core resistance per unit length will not be the same around the complete loop. Of course, the bridge will give a result that is electrically correct but it is necessary to make corrections to account for the different lengths of core of different cross-sectional area, so that the result will be correct, not only electrically as a comparison of two resistances but also in terms of core length.

The normal way of doing this on power cables is to convert lengths of the minority cross-sections to equivalent lengths of the majority cross-sections, or to convert the lengths of all other sections to equivalent lengths of the section in which the fault *probably* lies. This is best shown by example. Figure 3.14 represents a three-core copper medium voltage (MV) cable with sections of 150 and 120 mm² cores with lengths as shown.

A bridge balance gives a result of 75 per cent route length. Evidently, this location will be somewhere on the 120 mm² length, so the 345 m of 150 mm² cable should be converted to an equivalent length of 120 mm² cable.

The resistances of the two sizes are:

120 mm²   0.153 ohms/km
150 mm²   0.124 ohms/km

i.e. a ratio of $\dfrac{0.153}{0.124} = 1.2339$

Note that if there is no access to the relevant cable statistics, the inverse ratio of the two cross-sectional areas may be used, as resistance is inversely proportional to cross-sectional area. This, however, is not as accurate, i.e. in this case, 150/120 = 1.25. This does not apply, of course, when the two cable sections are of different material, e.g. copper and aluminium. In such cases, proper resistance figures must be used. Useful tables of cable statistics can be found in Chapter 14. Thus 345 m of 150 mm² cable is equivalent to

DIAGNOSIS AND PRE-LOCATION

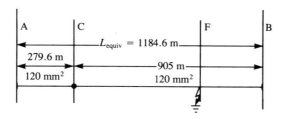

Fig. 3.15 Equivalent cable layout

$$345 \times \frac{1}{1.2339} = 279.6 \text{ m of } 120 \text{ mm}^2 \text{ cable}$$

i.e. a total equivalent length (L) of 120 mm² cable of 1184.6 m. This gives an equivalent cable layout as shown in Fig. 3.15. Therefore the location of 75 per cent lies at

$$1184.6 \times \frac{75}{100} = 888.45 \text{ m (in section CB as assumed)}$$

There are then two ways of locating this point overground:

1. As the cable in section CB is actually of 120 mm² cross-sectional area, the fault lies at

$$1184.6 - 888.45 = 296.15 \text{ m from end B}$$

2. The equivalent distance to fault (section AF) is 888.45 m which consists of 279.6 m (section AC) and 608.85 m (section CF). Therefore, the distance to fault from end A (section AF) is

$$608.85 + 345 \, (\textit{actual} \text{ length of AC}) = 953.85 \text{ m}$$

Cross-check: $953.85 + 296.15 = 1250$ m

Note that it is simpler and more convenient to carry out this type of calculation on *route length* as shown.

Calculations on a *loop length* basis will yield the same result, as follows. In the above example, the loop length equivalent diagram would be as shown in Fig. 3.16. In loop terms the bridge balances at 37.5 per cent. Therefore the fault distance is:

$$2L \times \frac{37.5}{100} = 2369.2 \times \frac{37.5}{100} = 888.45 \text{ m}$$

Considering this distance to be along the faulty core, f, from end A, the calculation then proceeds as in alternative 2 above.

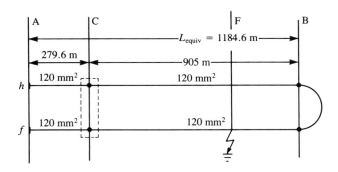

Fig. 3.16 Loop length equivalent diagram

CABLES WITH TEES

When carrying out bridge tests on cables with tee branches, it is sometimes necessary to make more than one test. Consider the teed cable shown in Fig. 3.17.

If there is a fault anywhere on the main cable, e.g. at F1, the test will proceed normally as if there were no tees. Providing the branches are healthy, their existence does not modify the main cable loop at all. Therefore they have no effect. However, if the fault is on a branch, e.g. at F2 on branch DQ, the initial test (with the strap at B) will give a location at the tee joint at D.

On obtaining a location at, or very near to, a tee joint, two possibilities exist: either the fault lies some way up the tee branch or the fault is indeed *at* the tee joint. When this happens, the strap should be removed from the main cable remote end B and put on at Q at the end of branch DQ.

The previous test will have been made with the cable length considered as L_1. Now, however, for the second test, the length used must be L_2, i.e.

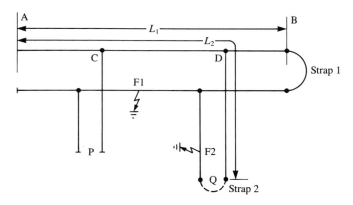

Fig. 3.17 Teed cable

Source to strap in all cases. The result, being a percentage of L_2, will clearly indicate whether the fault is at the tee joint or at some point along the tee branch, e.g. F2.

OTHER BRIDGE CONFIGURATIONS

There are cases where two healthy cores are available. These can be two of the other cores in the cable or cores in an auxiliary cable such as a pilot or control cable terminated in the same locations, A and B. If the cable is short, it may also be possible to run out an auxiliary cable overground. In such cases, the configuration shown in Fig. 3.18 may be used. This is called the Hilborn loop test.

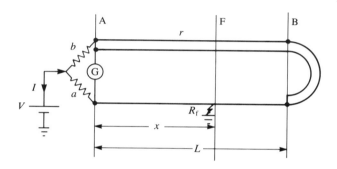

Fig. 3.18 Hilborn loop test

This is a powerful technique for the following reasons:

1 The resistances of the two auxiliary conductors are unimportant* as they appear in the internal resistance arms and the galvanometer circuit respectively. It follows that small connection resistances at end B are also unimportant – a most useful factor when the faulty core is of large cross-section.
2 In effect, the bridge is connected directly across the faulty core and the loop length is the route length. Errors are therefore smaller.

At balance:

$$x = \frac{a}{a+b+r} \times L$$

The Hilborn loop test is often used for the location of sheath faults as the

---

* And, if r is negligible with respect to the resistance of the internal arms of the bridge:

$$x = \frac{a}{a+b} \times L$$

34 UNDERGROUND CABLE FAULT LOCATION

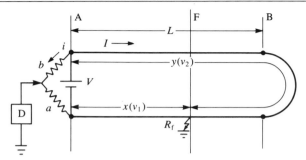

Fig. 3.19 Inverted loop test

cores of the cable are usually healthy and available for use as auxiliary connections.

A variation of the Murray loop test which is extremely effective on high resistance faults is the *inverted loop test*. The basic connection for this is shown in Fig. 3.19. From this it will be seen that the positions of the galvanometer (or null detector 'D') and the battery have been interchanged.

This means that the battery voltage V drives the current I around the external cable loop and the current i through the internal arms. No voltage greater than 6 V is required. This will produce currents in the cable loop of fractions of an ampere in telecommunications cables and roughly between one and one hundred amperes plus in power cables.

Therefore, on power cables, it is necessary to be able to connect and disconnect the battery quickly and conveniently. It also follows that, for power cable work, a heavy duty 6 V accumulator should be used which has sufficient capacity to drive the heavy currents around the cable loop and still maintain a potential difference (PD) not much below the nominal 6 V.

The arrangement is still a bridge and at balance:

$$\frac{a}{a+b} = \frac{x}{x+y}$$

and

$$x = \frac{a(x+y)}{a+b}$$

or

$$x = \frac{a}{a+b} \times 2L$$

Figure 3.20 shows a high resistance fault locator which operates on this principle.

DIAGNOSIS AND PRE-LOCATION   35

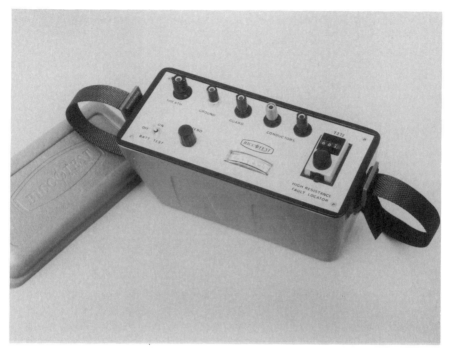

Fig. 3.20   High resistance fault locator (*source:* Biccotest Ltd, UK)

Such bridges are capable of locating faults of up to 200 megohms. However, the following points should be borne in mind when considering faults of such high resistances:

1. On most low voltage cables, insulation resistance (IR) readings of 200 megohms indicate a healthy cable!
2. If the cable is long enough to have a significant capacitance, a considerable time constant is involved when testing shunt resistance values in the order of hundreds of megohms.
3. Effects due to stray capacitance are observed.

and

4. Guard terminals should be connected to the core insulation at the A end connections to cancel out surface leakage.

In short, this is virtually a laboratory situation. In practice, the inverted loop test is extremely effective in locating faults of 0 to 10/20 megohms, which is satisfactory for most purposes.

The importance of the null detector arrangement must be appreciated

36 UNDERGROUND CABLE FAULT LOCATION

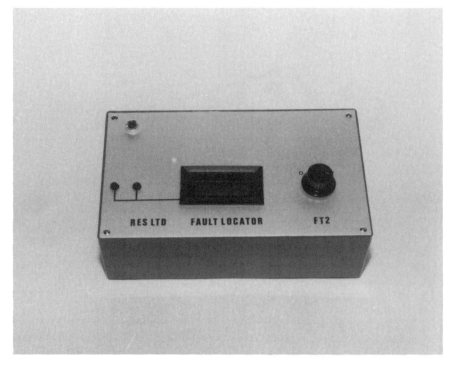

Fig. 3.21 The FT2 (*source:* Relay Engineering Services Ltd, UK)

also. That is to say, if high resistance faults are to be located, the zero voltage detector must have a very high input impedance so that the existence of megohms in the fault/return path is unimportant. This return path is effectively a voltmeter lead carrying a tiny negligible current.

The other important factor about the fault/return path is that, unlike the path in the Murray loop test, it is *not being stressed by an applied voltage*. It follows therefore that a successful test can only be carried out if there is a *fault return path*, even though it is of high resistance. A gap that a high voltage insulation resistance tester will break down will not be seen as a fault by the inverted loop tester. This again underlines the importance of carrying out the initial diagnosis with a low voltage ohmmeter.

A more modern instrument for this type of test, the FT2, from Relay Engineering Services Limited, is shown in Fig. 3.21. This is a computerized instrument capable of measuring and comparing voltages, and averaging and displaying the results. The built-in battery is switched on to the external loop as shown in Fig. 3.19 and the voltages v_1 and v_2 are sampled and the calculation:

$$\frac{v_1}{v_1 + v_2}$$

carried out. The polarity of the test is then reversed automatically such that

$$\frac{v_2}{v_1 + v_2}$$

is calculated. This means that the test has been carried out from both ends of the loop. The result can be displayed as a fraction, percentage or distance (if the cable length is first keyed in).

NEW INSTRUMENTS FOR 'BRIDGE' SITUATIONS

Refer back to the instruments in Figs 3.10 and 3.11 which show the CFL1 for telecommunications use and the FT1 for power or telecommunications cables respectively. As already stated, these are not bridges because they do not effect a balance. However, they are applied *in exactly the same way*, i.e. to a cable with one faulty and one (or more) healthy cores, looped at the remote end.

In the case of the CFL1, the general function of the instrument is shown in Fig. 3.22a and b. From this it can be seen that two tests are made by two switch selections (which are effected by depressing and releasing the single push button).

The first test, corresponding to Fig. 3.22a, connects a constant current d.c. source, I_{c1}, to the looped line which is in series with a reference resistance R_{ref1}. The voltages developed across the line and the reference resistance are fed to a ratiometric voltmeter and liquid crystal display (LCD) which displays the single line resistance in ohms. This is an absolute measurement which can be converted to units of distance using a conversion chart such as that in Fig. 3.23 on page 39.

The second test, shown in Fig. 3.22b, connects another constant current d.c. source, via the earth/fault path, to the section of cable core shown as R_{fault}. This is in series with another reference resistance, R_{ref2}. The two voltages developed across R_{fault} and R_{ref2} are again compared and a display produced of the absolute value of the core resistance between the source end and the point of fault. Therefore the fault distance can be derived from a conversion chart. Alternatively, the calculation:

$$\frac{R_{fault}}{R_{line}} \times 100$$

gives a location as a percentage of route length.

The maximum voltage available for the second test (resistance/distance

38 UNDERGROUND CABLE FAULT LOCATION

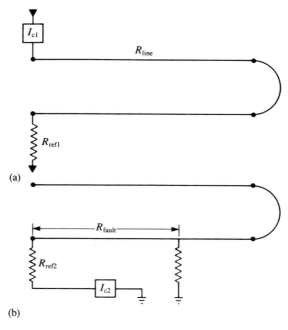

Fig. 3.22 The general function of the CFL1 for telecommunications use is shown here

to fault) is 190 V d.c. Therefore very high resistance faults will not allow enough test current to flow to enable a proper result to be obtained. The CFL1, however, is designed specifically for telecommunications applications in which a resistance greater than one megohm is not considered to be a fault condition. The display gives an indication when the fault resistance is above this value.

The FT1 (see Fig. 3.11) is also a small hand-held unit, but is used on power cables as well as telecommunications cables. Figure 3.24 shows the basic operation of the unit, from which it will be seen that a constant current high voltage d.c. source is switched alternately between points P and Q, the healthy and faulty cores respectively, and the voltages V_{p-q} and V_{q-p} are measured accurately and compared. The ratio of these voltages is the same as the ratio of R_x and R_y. The opto isolated microprocessor switches the relay, logs the multiple readings and then calculates, averages and displays the fault location as a distance or percentage depending on keypad input.

Table 3.2 on page 40 gives the test currents and results on several specimen equivalent cable lengths, showing reasonable location accuracies with very low test currents, even on heavy cables.

DIAGNOSIS AND PRE-LOCATION 39

MEGGER® Contact Fault Locator CFL1 Resistance/Distance Conversion Chart

Single Wire Copper Conductors

Resistance in ohms	Distance in metres for conductor diameter				
	0,32 mm	0,4 mm	0,5 mm	0,6 mm	0,63 mm	0,9 mm
0	0	0	0	0	0	0
0,1	0,5	0,7	1,2	1,7	1,8	3,7
0,2	1,0	1,5	2,3	3,3	3,7	7,5
0,3	1,4	2,2	3,5	5,0	5,5	11,2
0,4	1,9	3,0	4,6	6,7	7,3	15,0
0,5	2,4	3,7	5,8	8,3	9,2	18,7
0,6	2,8	4,4	6,9	10,0	11,0	22,5
0,7	3,3	5,2	8,1	11,6	12,8	26,2
0,8	3,8	5,9	9,2	13,3	14,7	30,0
0,9	4,3	6,7	10,4	15,0	16,5	33,7
1	4,7	7,4	11,6	16,6	18,3	37,4
2	9,5	14,8	23,1	33,3	36,7	74,8
3	14,2	22,2	34,7	49,9	55,0	112
4	18,9	29,6	46,2	66,5	73,4	150
5	23,7	37,0	57,8	83,2	91,7	187
6	28,4	44,4	69,3	99,8	110	225
7	33,1	51,7	80,9	116	128	262
8	37,9	59,1	92,4	133	147	299
9	42,6	66,5	104	150	165	337
10	47,3	73,9	116	166	183	374
20	94,6	148	231	333	367	748
30	142	222	347	499	550	1120
40	189	296	462	665	733	1500
50	237	370	578	832	917	1870
60	284	444	693	998	1100	2250
70	331	517	809	1160	1280	2620
80	378	591	924	1330	1470	2990
90	426	665	1040	1500	1650	3370
100	473	739	1160	1660	1830	3740
200	946	1480	2310	3330	3670	7480
300	1420	2220	3460	4990	5500	11200
400	1890	2960	4620	6650	7330	15000
500	2370	3700	5770	8320	9170	18700
600	2840	4440	6930	9980	11000	22500
700	3310	5170	8080	11600	12800	26200
800	3780	5910	9240	13300	14700	29900
900	4260	6650	10400	15000	16500	33700
1000	4730	7390	11500	16600	18300	37400

Single Wire Aluminium Conductors

Resistance in ohms	Distance in metres for conductor diameter				
	0,5 mm	0,6 mm	0,7 mm	0,8 mm	0,9 mm
0	0	0	0	0	0
0,1	0,7	1,1	1,4	1,9	2,4
0,2	1,5	2,1	2,8	3,7	4,7
0,3	2,2	3,1	4,3	5,6	7,1
0,4	2,9	4,2	5,7	7,5	9,4
0,5	3,6	5,2	7,1	9,3	11,8
0,6	4,4	6,3	8,6	11,2	14,1
0,7	5,1	7,3	10,0	13,0	16,5
0,8	5,8	8,4	11,4	14,9	18,9
0,9	6,5	9,4	12,8	16,8	21,2
1	7,3	10,5	14,3	18,6	23,6
2	14,5	20,9	28,5	37,2	47,1
3	21,8	31,4	42,8	55,9	70,7
4	29,1	41,9	57,0	74,5	94,3
5	36,4	52,4	71,3	93,1	118
6	43,6	62,8	85,5	112	141
7	50,9	73,3	99,8	130	165
8	58,2	83,8	114	149	189
9	65,5	94,3	128	168	212
10	72,7	105	143	186	236
20	145	209	285	372	471
30	218	314	428	559	707
40	291	419	570	745	942
50	364	524	713	931	1180
60	436	628	855	1120	1410
70	509	733	998	1300	1650
80	582	838	1140	1490	1890
90	655	942	1280	1680	2120
100	727	1050	1430	1860	2360
200	1450	2090	2850	3720	4710
300	2180	3140	4280	5590	7070
400	2910	4190	5700	7450	9420
500	3640	5240	7130	9310	11800
600	4360	6280	8550	11200	14100
700	5090	7330	9980	13000	16500
800	5820	8380	11400	14900	18900
900	6540	9420	12800	16800	21200
1000	7270	10500	14300	18600	23600

MEGGER INSTRUMENTS LIMITED
Archcliffe Road, Dover, Kent CT17 9EN, England.
Tel: 0304 202620 Fax: 0304 207342 Telex: 96283 Avomeg G
MEGGER is the registered Trade Mark of MEGGER INSTRUMENTS LIMITED
Part No. 5173-015 Edition 1 Printed in England SL/0,5k/3T

Fig. 3.23 The Megger contact fault locator CFL1 resistance/distance conversion chart (*source*: AVO International, UK)

TELECOMMUNICATIONS BRIDGES AND CONTACT FAULT LOCATORS
Bridges designed specifically for telecommunications work normally measure resistance to the fault directly in ohms, e.g. the instruments shown

Table 3.2 Test currents and results

Route length (m)	Cross-section (mm²)	Loop resistance (Ω)	Current (mA)	Location Test (%)	Actual (%)
77	2.5	1.1	2.77	45.00	45.00
424	300	0.05	2.77	32.48	33.33
1415	185	0.275	0.46	40.37	40.00

in Figs 3.8 and 3.10. The loop resistance can also be measured. Once the resistance to fault is known, it is normal to use conversion charts to find the distance this resistance represents along the particular type of cable concerned. An example of such a chart is given in Fig. 3.23, from which it can be deduced that a resistance to fault of, for instance, 100 ohms represents a distance of 739 m on a 0.4 mm copper pair.

As absolute values of resistance per unit length are being used, it is essential that the correct conversion chart is used for the material and size of conductor involved. If no chart is available and the route length is known accurately, then the result is given by

$$\frac{\text{Resistance to fault}}{\text{Loop resistance}} \times \text{loop length}$$

It is implicit in any measurement of resistance to fault that all calculations are in terms of *loop resistance* or *length*, i.e. resistance is being measured *along one leg of a pair*. Thus the derivation of a fault distance from a

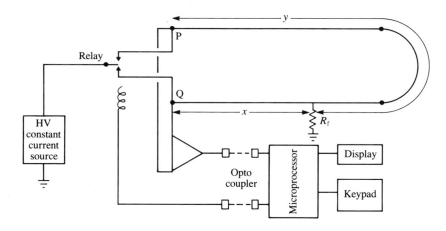

Fig. 3.24 The basic operation of the FT1

DIAGNOSIS AND PRE-LOCATION 41

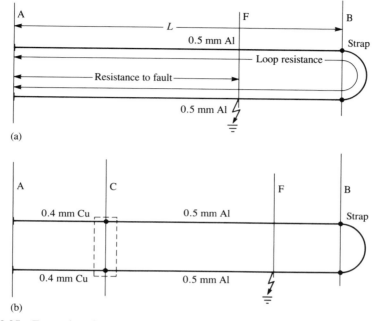

Fig. 3.25 Examples of the calculation of fault distance from resistance to fault

resistance to fault using look-up tables is a straightforward matter not involving any adjustment for *loop* or *route length*.

The following examples relate to Fig. 3.25a and b respectively:

(a) Cable: 0.5 mm Al, length 195 m
    Loop resistance reading: 53.64 ohms
    Resistance to fault reading: 18.3 ohms

From the conversion chart (Fig. 3.23), the distance to fault is:

Resistance (ohms)	*Distance* (m)
10.0	72.7
8.0	58.2
0.3	2.2
18.3	133.1

Alternatively, worked as a percentage, this is

$$\frac{\text{Resistance to fault}}{\text{Loop resistance}} \times \text{loop length} = \frac{18.3}{53.64} \times 390$$

$$= 133.05 \text{ m}$$

(b) Cable: 0.4 mm Cu, length   84 m
      plus   0.5 mm Al, length   228 m

From the conversion chart it is clear that the fault is beyond the joint at C, i.e. in the 0.5 mm aluminium section, because a resistance to fault of 33.53 ohms gives 247.78 m on 0.4 mm copper cable. Therefore, the resistance of the 0.4 mm copper section up to C should be calculated as follows:

$$84 \text{ m at } 0.135\,32 \text{ ohms/m} = 11.37 \text{ ohms}$$

The remaining resistance from C to the fault at F is then:

$$33.53 \text{ ohms} - 11.37 \text{ ohms} = 22.16 \text{ ohms}$$

along the 0.5 mm aluminium section. This is equivalent to 161.1 m. Therefore, the distance to the fault *overground* is

$$84 \text{ m} + 161.1 \text{ m} = 245.1 \text{ m}$$

There are intelligent telecommunications bridges that have extra facilities including compensation for conductor temperatures, readout directly in metres or feet and the ability to carry out locations on cables with up to four sections of different cross-section. They can also locate high resistance faults of up to *ca.* 10 megohms.

Another type of telecommunications fault is the 'battery fault' or 'foreign battery'. This occurs when there is contact or partial contact at the point of fault with the -48 V potential of the exchange/central office battery. In the bridge configuration this appears as a d.c. source of up to 48 V in series with the fault/return path.

In the Murray loop connection this presents no major problem as enough test voltage is, or can be made, available to 'buck' this external d.c. source and drive current through the fault in order to obtain a balance. With the basic inverted loop test, however, there is no test voltage being applied to the fault/return path and the location of battery faults should not be attempted with this test, unless the instrument being used is a more sophisticated version with the capability of nulling out this external voltage.

3.4   Pulse echo (also called time domain reflectometry)

In the pulse echo method of pre-location, a pulse is launched into a cable from one end. This pulse is completely or partly reflected by any impedance mismatch it encounters and, because the velocity of propagation is taken to be constant for a given cable, the time taken for the pulse to return to the source is a measure of the distance to the mismatch. A mismatch can be a shunt path (e.g. a short circuit fault) or a series path (e.g. an open circuit fault or end).

The velocity of propagation, V, depends on the cable dielectric and is defined as

$$\frac{v_s}{\sqrt{\varepsilon}}$$

where
    v_s = velocity of light in free space (300 m/μs or 984 ft/μs)
    ε = relative permittivity of the dielectric

Table 3.3 is a list of typical velocities of propagation for different dielectrics.

Table 3.3 Typical velocities of propagation for different dielectrics

Dielectric	*Velocity of propagation*
Impregnated paper	150–171
Dry paper	216–264
PE	Approx. 200
XLPE	156–174
PVC	152–175
PTFE	Approx. 213
Air	Approx. 282

These are approximate as values can vary due to differences in a particular type of insulation, temperature and the ageing of the insulation. Such values can be used if all else fails, but it is usually possible to use an accurate value for V derived empirically from a known length of the same cable or to use a proportional method as outlined later with reference to Fig. 3.32.

Some manufacturers quote a velocity factor instead of velocity. This is simply the ratio of the velocity to the velocity of light, e.g. for a velocity of propagation of 160 m/μs, the velocity factor is

$$\frac{160}{300} = 0.533$$

Any pulse travelling in a cable is a high frequency phenomenon which sees a cable of infinite length before it with an impedance of Z_0, the characteristic impedance defined as

$$z_o = \sqrt{\frac{L}{C}}$$

Values for power cables vary between 15 and 80 ohms and those for telecommunications cables between 100 and 1500 ohms. Common average values used for these are 40 and 600 ohms respectively.

The reflection factor, r, is a function of the fault resistance R_f and the characteristic impedance of the cable concerned, Z_0. For shunt faults it is

$$r = \frac{-Z_0}{2R_f + Z_0}$$

and for series faults

$$r = \frac{R_f}{2Z_0 + R_f}$$

To get an idea of the magnitudes of trace features for various faults and cable features it is useful to make a few specimen calculations of reflection factors, as shown in Table 3.4a and b for power cables and in Table 3.5a and b for telecommunications cables. It can be seen from these tables that approximate limiting values of fault resistance for shunt faults are 400 ohms for power cables and 6000 ohms for telecommunications cables. In other words, faults with resistances greater than $10 \times Z_0$ are unlikely to produce recognizable trace features.

Moreover, in practice, this should be taken as an absolute maximum because the calculations presuppose that the pulse *as transmitted* is being reflected. In fact, the losses and minor mismatches in the cable cause the pulse to be progressively attenuated and flattened the further it travels. Therefore, it may well be possible to recognize a trace feature from a mismatch of $10 \times Z_0$ at, say, 100 m from the source, but a similar mismatch at 1000 m will certainly not be visible.

Figure 3.26 is a graph of percentage reflection against fault resistance for pulses into a 40 ohm power cable for shunt and series faults, while Fig. 3.27 shows the equivalent curves for a 600 ohm telecommunications cable. It should be noted that shunt faults produce *negative* trace features and series faults *positive* ones.

Thus a zero ohm short circuit will give a trace feature as shown at F in Fig. 3.28. The leading edge is steep and the trailing edge less so. A zero ohm fault is a *total* mismatch and *all* the pulse energy is reflected. This leaves no pulse to carry on past F and therefore the end of the cable does not show.

Series faults and ends produce a *positive* trace feature, as shown in Fig. 3.29. A 'clean' open circuit is also a total mismatch; therefore all the pulse energy is again reflected back to source leaving no pulse to travel further. The feature shown at end A is due to the mismatch between the output circuit of the PE set/TDR (including test leads) and the cable. This feature can be minimized with the 'matching/balance' adjustments available on some sets.

DIAGNOSIS AND PRE-LOCATION 45

Table 3.4 Power cables
(a) Shunt faults: $Z_0 = 40$ ohms, $r = -Z_0/(2R_f + Z_0) \times 100$

Fault resistance (Ω)	Calculation	Reflection factor (%)
1	$\dfrac{-40}{2+40} \times 100$	-95
10	$\dfrac{-40}{20+40} \times 100$	-67
40	$\dfrac{-40}{80+40} \times 100$	-33
100	$\dfrac{-40}{200+40} \times 100$	-17
400	$\dfrac{-40}{800+40} \times 100$	-4.8

(b) Series faults: $Z_0 = 40$ ohms, $r = R_f/(2Z_0 + R_f) \times 100$

Fault resistance (Ω)	Calculation	Reflection factor (%)
5m	$\dfrac{5\,000\,000}{80 + 5\,000\,000} \times 100$	Approx. 100
100k	$\dfrac{100\,000}{80 + 100\,000} \times 100$	Approx. 100
1000	$\dfrac{1000}{80 + 1000} \times 100$	93
100	$\dfrac{100}{80 + 100} \times 100$	55
40	$\dfrac{40}{80 + 40} \times 100$	33
10	$\dfrac{10}{80 + 10} \times 100$	11
1	$\dfrac{1}{80 + 1} \times 100$	1.2

In the case of a fault with resistance greater than zero ohms, only a portion of the pulse energy is reflected (see Tables 3.4 and 3.5 and Figs 3.26 and 3.27). This leaves some energy left to cause reflections from subsequent mismatches.

Figure 3.30a shows the trace from a low resistance shunt fault on which features for both the fault and the end are visible. Note that the end feature is quite small. If the fault is of a higher resistance, the fault feature will, of course, be smaller, because less energy has been reflected back from it, while more energy travels on to be reflected from the end to give a larger end feature. This is shown in Fig. 3.30b.

MEASUREMENTS

All measurements are made to the *start* of the trace feature as shown in

46 UNDERGROUND CABLE FAULT LOCATION

Table 3.5 Telecommunications cables
(a) Shunt faults: $Z_0 = 600$ ohms, $r = -Z_0/(2R_f + Z_0) \times 100$

Fault resistance (Ω)	Calculation	Reflection factor (%)
1	$\dfrac{-600}{2 + 600} \times 100$	Approx. -100
10	$\dfrac{-600}{20 + 600} \times 100$	-97
100	$\dfrac{-600}{200 + 600} \times 100$	-75
600	$\dfrac{-600}{1200 + 600} \times 100$	-33
1000	$\dfrac{-600}{2000 + 600} \times 100$	-23
6000	$\dfrac{-600}{12\,000 + 600} \times 100$	-4.8

(b) Series faults: $Z_0 = 600$ ohms, $r = R_f/(2Z_0 + R_f) \times 100$

Fault resistance (Ω)	Calculation	Reflection factor (%)
5m	$\dfrac{5\,000\,000}{1200 + 5\,000\,000} \times 100$	Approx. 100
100k	$\dfrac{100\,000}{1200 + 100\,000} \times 100$	99
1000	$\dfrac{1000}{1200 + 1000} \times 100$	45
600	$\dfrac{600}{1200 + 600} \times 100$	33
100	$\dfrac{100}{1200 + 100} \times 100$	7.7
10	$\dfrac{10}{1200 + 10} \times 100$	0.01

Fig. 3.31, i.e. to the exact point where the trace leaves the x axis. In the early days of PE/TDR testing, measurements were made by using the graticule markings and a *shift* control which moved the trace to the left. Nowadays, all PE/TDR instruments utilize some sort of cursor (vertical line, bright-up trace or light spot), which is moved to the selected position on the trace.

This operation also gives a digital readout of the 'distance' to the fault in microseconds. The actual distance to the fault is then calculated by multiplying the time of flight in microseconds by the velocity of propagation in metres or feet per microsecond. Because the pulse travels to the fault and *back*, it is convenient to use the value V/2 so that the actual distance to fault along the cable is derived. Most instruments now incorporate a facility for inputting

DIAGNOSIS AND PRE-LOCATION   47

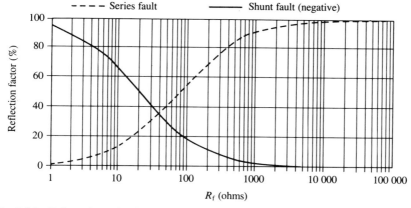

Fig. 3.26 Pulse echo reflection factors (power cables)

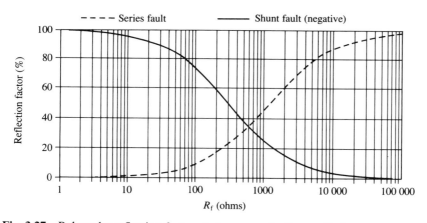

Fig. 3.27 Pulse echo reflection factors (telecommunications cables)

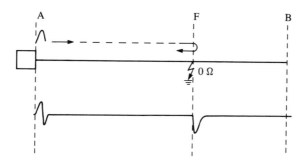

Fig. 3.28 A zero ohm short circuit gives a trace feature as shown at F.

48 UNDERGROUND CABLE FAULT LOCATION

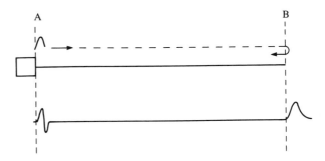

Fig. 3.29 A positive trace feature produced by series faults and ends

the velocity of propagation so that the calculation is made internally with the actual distance displayed in units, usually metres or feet, which can often be pre-selected.

PE/TDR is inherently less accurate than bridge methods because of:

1 The difficulty often experienced in choosing the exact point to measure to.
2 The values of velocity of propagation being only approximate and,

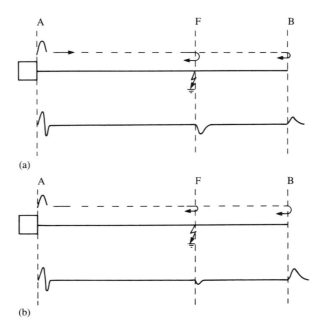

Fig. 3.30 Traces from shunt faults, (a) low resistance and (b) higher resistance

DIAGNOSIS AND PRE-LOCATION 49

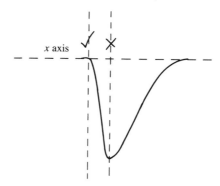

Fig. 3.31 Measurements made to the start of the trace feature

contrary to popular belief, the velocity of propagation changing with the varying frequency content of the pulse which is distorted as it travels along the cable. This effect is less marked in pulse echo tests than in transient tests (see Fig. 3.47 in the section on 'digital fault locator').

3 Minor variations in velocity of propagation over the route length when different cables are jointed together.

Therefore great care should be taken over all aspects of measurement. Figure 3.32 shows an example of a fault at 3 μs on a cable whose end is at 5 μs. If the $V/2$ for the cable concerned is known to be 80 m/μs, the fault lies at

$$3 \, \mu s \times 80 \, m/\mu s = 240 \, m$$

should this value be uncertain, however, the result could well give a location between, say, 225 and 255 m – up to 6 per cent either side of the true location!

It is essential, therefore, to choose one of three options:

1 Use a trustworthy value for $V/2$.
2 Employ a proportional method based on the *correct* cable length.
3 Confirm the $V/2$ empirically for the particular type of cable concerned.

Using option 2, the location of the fault shown in Fig. 3.32 is simply:

$$\frac{3}{5} \times \text{cable length}$$

Option 3 necessitates carrying out a measurement to the end of a reasonable length of cable of exactly the same type of cable as the one in question and of *known* length. This is often quite easy to do on new cable drums in a depot. If records of previous fault locations have been kept, $V/2$ can be

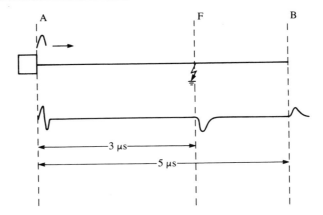

Fig. 3.32 Example of a fault at 3 µs on a cable whose end is at 5 µs

derived accurately from measurements already carried out on the same type and manufacture of cable on the same or a similar network.

TRACES FOUND IN PRACTICE

A trace often encountered is the one shown in Fig. 3.33. The features x, y and z are simply re-reflections of the strong *end* feature at B. Such re-reflections are usually easy to recognize because they are equidistant and diminishing, the time interval t between them being the same. This can be ascertained by choosing a long enough range such that all are visible and then measuring to each feature.

It is *always* a good idea to select a range longer than the presumed length of the cable just in case:

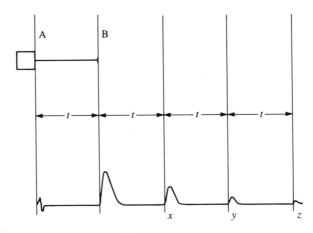

Fig. 3.33 Re-reflections of the end feature

1 The cable is significantly longer than was thought.
2 There is uncertainty about the 'end' feature displayed.

or

3 An end feature is evident at a point significantly further than the known length of the cable. This denotes the existence of an unknown tee branch near the end of the cable.

The fact that successive features diminish in amplitude is due to losses in the cable, the pulse energy being steadily dissipated the further the pulse travels.

The magnitude of these features can be reduced by better matching between the test set and the cable. Many test sets have a facility for adjusting this. A small start feature together with small or non-existent re-reflections of the end are indicative of good matching.

As well as faults and ends, other cable features can give rise to reflections. These include joints, changes of cable cross-section, core splits and re-splits and waterlogged zones, all of which are minor mismatches. Tee joints and ends of tee branches are more significant mismatches which produce trace features of greater magnitude. The theoretical reflections resulting from these are as shown on the upper trace in Fig. 3.34.

Taking these cable features in turn, the straight joint at u produces a small trace feature which comprises a small positive excursion followed by a small negative one. This is because, as the cores within the joint separate, the inductance, L, increases and the capacitance, C, decreases. Thus the characteristic impedance, Z_o, which is $\sqrt{L/C}$, as stated earlier, increases, giving rise to a positive reflection.

Conversely, as the cores come together to exit the joint, the characteristic impedance decreases, producing a negative reflection. Of course, the joint itself is probably less than 1 m long, so these indications mix into a very small feature, as shown at u on the upper trace. This is particularly true when the cable sections either side of the joint are of the same cable. In practice such a joint will be totally invisible unless it is very close to the test end. It is also undetectable when it is physically close to a major cable feature such as a fault or open circuit.

More significant trace features are produced by joints connecting cables with different characteristic impedances. Looking from the test end, when the cable after the joint has a *smaller* characteristic impedance than the one before, a *negative* feature is produced. The opposite change, *smaller* to *greater* characteristic impedance, results in a *positive* trace feature.

When the pulse arrives at the tee joint at v, the energy splits up, some carrying on towards end B, some travelling up the tee branch to w and

52 UNDERGROUND CABLE FAULT LOCATION

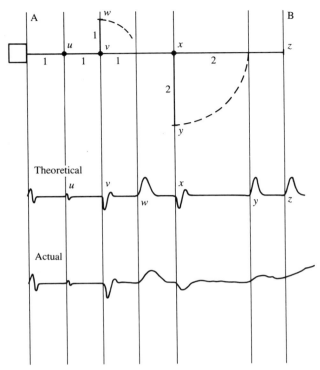

Fig. 3.34 Theoretical and actual traces for a cable with straight and tee joints

some being reflected back to source, giving rise to the feature shown at v on the trace. This is a sizeable negative feature with a smaller positive overshoot depending on the length of the tee branch. Great care must be taken not to confuse this type of feature with that produced by a shunt fault (which recovers straight on to the x axis).

The reflection factor for a tee joint is 33 per cent, being the same as that for a shunt fault of resistance equal to the characteristic impedance of the cable (refer to Tables 3.4 and 3.5 and Figs 3.26 and 3.27).

The portion of the pulse that travels up the tee branch is reflected from the end of the tee branch at w and produces an 'open circuit' type feature, as shown at w on the upper trace. This is because measurements along the x axis of a PE/TDR display are in terms of the 'running time' of the pulse, the only criterion for the mutual displacement of trace features being *time lapse*. Therefore the 'tee branch' pulse portion reaches the point w at the same time as the 'main cable' pulse portion. Though on different parts of the cable, the pulses have run for the same length of time.

In the example shown in Fig. 3.34, the numbers represent units of time lapsed, say microseconds. Thus the time taken for a pulse to travel from the tee joint at v to the end of the tee at w is 1 μs, the same as a pulse travelling along the main cable for 1 μs.

Similarly, the pulse continuing towards end B splits again at the tee joint x, part going up the branch to y and part travelling on to z (end B). In this case the branch is longer and the travelling time is shown as 2 μs so that the pulse reaches the end of the branch, y, to give a positive feature at y on the trace.

The feature at z is, of course, the end of the main cable. It should always be borne in mind that, if such a branch is longer, say 3 μs in this case, the branch end feature will be beyond the main cable end feature.

Unfortunately, the 'picture' is not like this in practice! As mentioned earlier, the losses in the cable reduce the energy of the pulse the further it travels. Also, and more importantly, each mismatch encountered reflects energy back so that successive cable features are hit by smaller and smaller pulses and the reflections received back at source are themselves smaller.

Add to this the fact that the pulse is progressively broadened and made less sharp the further it travels, and it will be clear that, in practice, trace features further along a cable will be smaller in magnitude and more rounded in form. Indeed, at some point along the trace, there will be no identifiable discrete features and the latter part of the trace will be meaningless. In practice, the cable depicted in Fig. 3.34 will produce a trace somewhat like the 'actual' trace shown.

The matter of pulse width is very important. Narrow pulses give rise to very sharp trace features which are easy to measure to, but they are quickly attenuated and are therefore only useful over short distances. Wide pulses, on the other hand, produce wider and more rounded trace features whose 'start points' are more difficult to identify. However, they are not so quickly attenuated and therefore travel further. The half-height pulse widths produced by commercially available PE sets/TDRs vary from 2 ns (approximately equivalent to 0.16m) to several microseconds (equivalent to a few hundred metres).

TOWARDS BETTER MEASUREMENT

When a fault feature is clear and unambiguous, measurements can be made as described so far and accuracies better than 0.5 per cent can be obtained. All too often, however, it is not certain whether a particular feature is or is not a fault, and in such cases other means have to be employed to make a successful pre-location. There are several ways of doing this which involve using different techniques or electronic facilities incorporated in the instrument, or, more usually, both.

A very useful approach is the *comparison* of traces from healthy and

54 UNDERGROUND CABLE FAULT LOCATION

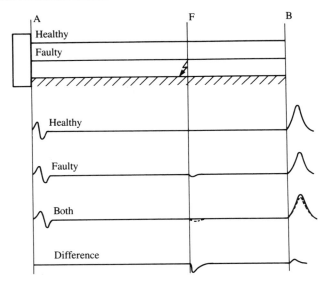

Fig. 3.35 The comparison of traces from healthy and faulty cores or pairs

faulty cores or pairs. Figure 3.35 shows a situation in which a fault of, say, several hundred ohms is extremely difficult to identify as a discrete feature. However, if a healthy core is viewed, a significant difference can be detected at F.

Obviously it would be better if both these traces could be viewed at the same time. Two electronic facilities are available to accomplish this. One is a comparison switch which quickly switches one pole of the output between the healthy and the faulty core so that, due to persistence of vision and the retentivity of the luminous screen coating, the two traces appear to be superimposed on the screen. It is then an easy matter to measure to the point at which the traces diverge. Such a vibrator may be built into a particular instrument or the user can construct a simple battery operated device and connect it to the output. Note that even if the available set has a comparison facility, it is not a bad idea to make up such a device so that it can be connected to a cable termination to positively identify the remote end on the screen.

Instruments fitted with phase or pair comparison switches have existed for many years but recently a more modern facility has become generally available – the incorporation of one or more *memories*. This enables the trace from one core or pair to be stored in memory while another is viewed in real time. The two can then be superimposed, mutually displaced, added or subtracted. In fact, the 'difference' display as shown in Fig. 3.35 is most indicative.

DIAGNOSIS AND PRE-LOCATION 55

This development, the provision of memory, enables an extremely powerful technique to be easily exploited, i.e. the comparison of traces of a fault *before and after* treatment. Treatment is any action taken by the operator to modify the nature of the fault. This is usually done by burn-down (see a later section on 'treatment of faults'), but can also be accomplished by drying the fault out or by changing its condition by surging into the cable (see Section 2.4 and Chapter 4) or by re-energizing the cable. The very important fact of the matter is that, *providing a trace has been put into memory initially, any change that can be produced by whatever means will result in the location of the fault.*

Most PE/TDR instruments can be used on both power and telecommunications cables but there are some which either incorporate features more applicable to telecommunications cables or which are specifically designed for telecommunications use. The main facility of these instruments is the provision of several bipolar outputs with which healthy and faulty pairs may be viewed and compared. A healthy pair can also be connected as a 'balance' to cancel out the features, e.g. joints, which are common to both pairs.

Two outputs can also be used to find cross-talk faults, whereby the outgoing pulse is transmitted on one pair and received back on another pair via the coupling at the fault (see Fig. 3.36).

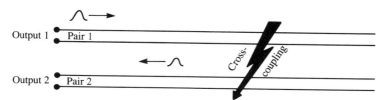

Fig. 3.36 Cross-talk fault

TYPES OF PULSE ECHO/TIME DOMAIN REFLECTOMETRY INSTRUMENT

Many instruments are currently available. They range from small low cost units, some of which have liquid crystal displays instead of the usual cathode ray oscilloscope (CRO), to larger more expensive ones which incorporate every refinement and facility. Notwithstanding the ever-present budgetary constraints, it is always wise to choose carefully, bearing in mind the facilities offered by the instrument related to the field requirements.

Figure 3.37 shows some of the different types of PE set/TDR. Broadly speaking there are three bands of requirements:

- Occasional use – short cables – no tees – faults mainly breaks and shorts – no requirement for interpretation – budget limitations

56 UNDERGROUND CABLE FAULT LOCATION

- Frequent use – long and short cables – complex networks – ability to interpret traces – full complement of other instruments available – no severe budget limitations
- Generally as for frequent use above, but for constant use by experts on all types of system – probably for housing in a test van with a high incidence of transient tests being required – no cost limit for right equipment

The groups of instruments that best match these requirements are:

1. A lower cost group with a small CRO or LCD and the essential facilities of range and velocity of propagation selection, measurement cursor and result display/readout. Typically, such a set would be used for initial

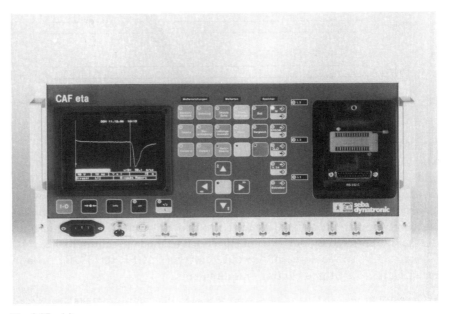

Fig. 3.37 (a)

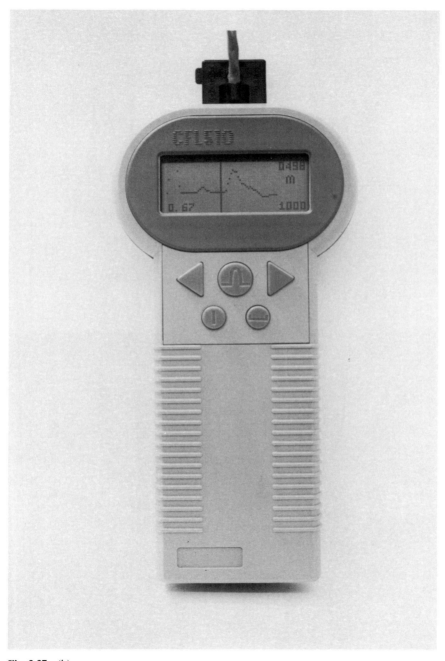

Fig. 3.37 (b)

58 UNDERGROUND CABLE FAULT LOCATION

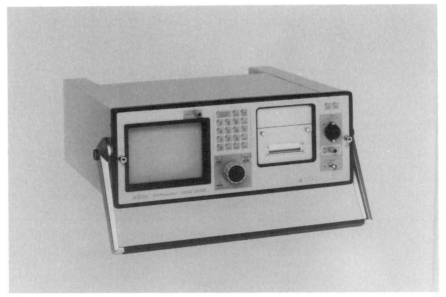

Fig. 3.37 (c)

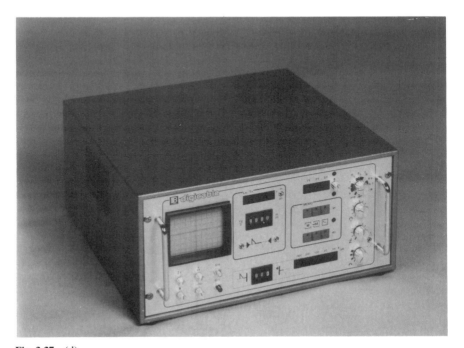

Fig. 3.37 (d)

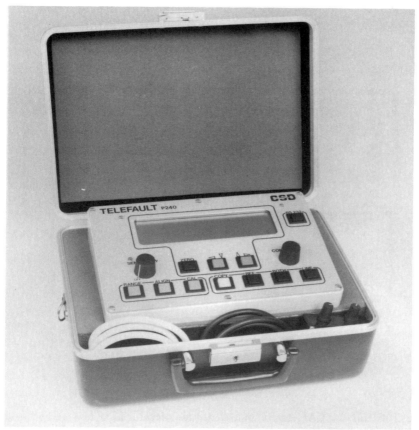

Fig. 3.37 (e)

diagnosis/location by telecommunications faultsmen/jointers or engineers installing short runs of power cable. These operators would have a resistance bridge or contact fault locator for medium and high resistance contact faults, but no equipment for burning-down or pin-pointing a fault.
2 The medium cost group of instruments which are portable but which incorporate almost all the refined facilities, e.g. matching, balance and sensitivity controls, range selection, dual outputs, core comparison, memories, pulse width selection, line cursors, readout in different units, internal/external supply choices, hard copy or printer/plotter interface. This type of set may or may not have an input for externally derived signals.
3 The higher cost group made up of sets designed to display and analyse externally derived transients as well as the self-generated pulses used for

60 UNDERGROUND CABLE FAULT LOCATION

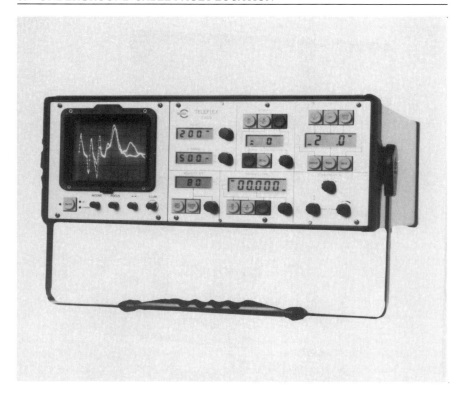

Fig. 3.37 (f)

Fig. 3.37 (a) The CAF eta computer aided fault locator (*source:* Seba Dynatronic GmbH, Germany). (b) The T510 hand-held pulse echo set (*source:* Biccotest Ltd, UK). (c) The IRG 300 computer aided pulse echo set/monitor (*source:* Baur Prüf- und Messtechnik GmbH & Co. KG, Austria). (d) The DIGICABLE microprocessor controlled reflectometer (*source:* Balteau, S.A. Belgium). (e) The TELEFAULT P240 (*source:* Hathaway Instruments Ltd, UK). (f) The TELEFLEX T 01/6 pulse echo set with integrated transient recorder (*source:* Hagenuk GmbH, Germany)

PE/TDR location. These sets are usually larger mains-powered units suitable for installation in test vehicles and capable of carrying out all PE/TDR, transient and arc reflection tests and measurements. They can also have features such as multiple memories, data storage, standard interface, menu drive, etc.

TREATMENT OF FAULTS

The normal way to change the condition of a fault is to *burn it down*. As stated earlier, it is very bad practice to use burn-down as a routine approach. However, there are many occasions when it is advisable or

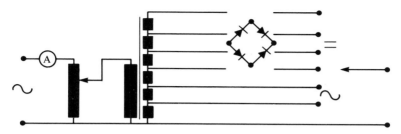

Fig. 3.38 Schematic representation of a typical burn-down set

necessary to try to reduce the resistance of the fault to a value that will enable a pre-location to be made. There are, of course, pre-location methods involving the application of high voltage where burn-down is never used, but these methods are not always available.

Burn-down requires three things:

- A high enough voltage to break the fault down
- A high enough current to carbonize the fault path
- The means to control these

Modern burn-down sets are designed specifically with these in mind. There is no escaping the fact that combination of the first two properties above means *power*. This in turn means *weight*; therefore the instruments tend to be quite heavy.

Figure 3.38 is a schematic representation of a typical burn-down set which shows it to be basically a transformer producing high voltage (HV) a.c. with voltage and power control in the primary. Several HV tappings are then rectified to produce HV d.c. outputs. There are also usually several a.c. outputs as shown to provide heavy current at low voltage.

Burn-down sets fall into two main categories:

1. Portable sets rated at around 1 or 2 kVA with d.c. outputs up to about 3 kV. These may or may not have a.c. outputs.
2. Sets rated up to approximately 10 kVA with a maximum d.c. output in the region of 15 kV, and several a.c. outputs for heavy burn-down with currents approaching 100 A.

Because of the power requirements all such units are mains powered. The sets in the first group are for burn-down on low voltage or telecommunications and control cables. They can also be used for pin-pointing ground contact faults such as 'sheath faults' (see Sec. 4.3). Instruments in the second group are used for burn-down on MV and HV power cables and would normally be mounted in test vans or trailers.

62 UNDERGROUND CABLE FAULT LOCATION

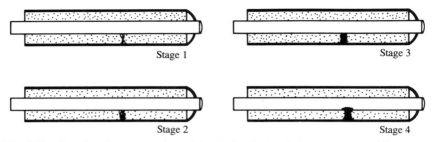

Fig. 3.39 Burning down a weak path in the insulation between two metal lines

All units have current indication, usually on a simple ammeter in the primary circuit, as exact current measurement is not necessary. A general indication of current throughout is all that is required. Some models also have a voltmeter. Output voltage is normally regulated with a variable transformer in the primary.

However, control of current or power drawn from the unit is effected in a variety of ways. It will be appreciated that a burn-down set cannot be allowed to 'trip out' if short circuited. Its very function is to allow a controlled current to pass through the fault in order to burn it down to a lower resistance value. Thermal overload trip and/or alarm are often incorporated just in case damage may be imminent after prolonged periods of use on full output.

One of the methods of controlling this current is to use a thyristor control circuit in the primary side. This enables the power being fed through to be adjusted quite easily with a potentiometer. This method is often incorporated in the smaller sets and permits a very fine control of throughput. Its main disadvantage is that the control circuit contains several electronic components which may fail.

A simpler system often employed in the larger sets uses no components at all. The approach referred to is the use of a 'short circuit proof' transformer as the main LV/HV transformer, i.e. a 'lossy' transformer with a high percentage impedance. This simple, self-limiting arrangement makes for instruments that are robust, reliable and function very satisfactorily in the field.

As shown in Fig. 3.39, the process of burning down a weak path in the insulation between two metal lines goes through several stages. When burning down it is normal to select a high d.c. voltage range, the maximum voltage of which lies below the test voltage of the cable, but which will nevertheless be capable of stressing the fault to breakdown. The voltage is raised slowly until the fault arcs over and current flows. The ammeter is liable to kick spasmodically as the current flow starts and stops. This happens because, initially, there is a high d.c. potential across the fault and no current flowing. Breakdown then occurs and current flows. This, however, causes the voltage to sag. The stress across the path is therefore decreased and the current stops.

In turn this allows the voltage to rise and the arc is re-established. The voltage control should be raised until this 'hunting' quickens up and the arc is maintained. This is stage 1. Only a few milliamps are flowing and sparking is occurring along tiny paths between the two metal lines, these paths breaking and being re-established and new paths being formed. Now the carbonization of the dielectric is beginning but, if the voltage is removed, the fault resistance would not have been appreciably lowered. There would not yet be enough carbon created and the path would have many gaps.

At this point in the proceedings an attempt can be made to switch down to the next d.c. voltage range. If the burn ceases, the first range should be quickly re-selected to establish the arc again. However, if current continues to flow at the lower range, the burn-down is progressing well and, as the applied voltage is now lower, the current is higher – possibly several hundred milliamps. Thus stage 2 is reached, where the several carbon paths are being consolidated and thickened. It may well be that the path will hold up if the burn is now stopped. However, it is more likely that the fragile threads of carbon will crack on cooling.

It is normally necessary therefore to move on to stage 3 by switching down to a yet lower voltage range, thereby increasing the current still further. This should build up the carbon bridge quite rapidly as considerable power is being dissipated. The following questions now arise:

1 How long should this stage be continued?
2 Will the carbon bridge break if the burn is stopped?
3 Has the burn gone on too long, i.e. is the carbon bridge now too solid?

The purpose of burning down the fault to a lower resistance is usually to enable the fault to be 'seen' on a PE set/TDR. For this, the fault resistance should preferably be a few ohms. However, it is *absolutely essential* that, after the fault has been successfully pre-located, the application of repeated surges from a shock wave generator should break up the carbon bridge and change the fault resistance from low to medium or high. Most of the energy from the shock wave generator will then be dissipated explosively at the fault, enabling it to be pin-pointed by the acoustic method. It is important to note that surging into a zero ohm fault produces *no sound at all*, i.e. no energy can be dissipated in a zero ohm load.

From the above questions and statements it is clear that some degree of expertise is required in the use of burn-down equipment. Though this is certainly the case, the following points should help the operator. While the burn is in its early stages, the burn-down set can seem quite noisy due to the rapid switching on and off, or 'hunting' as already explained. At this stage it is too early to switch off.

There usually comes a point, however, at which the instrument suddenly settles down to a quiet, smugly self-complacent hum! This signifies that the fault path is a constant low resistance. If the burn is now continued for 5 or 10 minutes there is every likelihood that the path will become too solid. Therefore, unless this is the intention (see page 67), the burn should be terminated after, say, 30 seconds to 2 minutes.

At this juncture, as soon as the output connections have been safely 'dumped' to earth, the resistance of the fault must be checked *very quickly* with a low voltage ohmmeter. If this is left connected it will be possible to see variations in resistance as the carbon path is stressed as it cools. The resistance of the fault must be monitored at all stages of the fault location so that logical decisions can be made about how to proceed. However, *directly after* the ohmmeter registers a low fault resistance, the PE set/TDR must be connected *immediately*, so that the fault feature may be quickly identified on the trace, just in case the fault resistance goes high again.

Consider the following often enacted scenario:

1 A high resistance fault cannot be seen on the trace.
2 It is successfully burnt down.
3 The fault resistance is checked. It is low.
4 While connections are being changed, *the fault resistance goes high again*.
5 The PE set/TDR is reconnected.

Result: no sign of the fault. *Again, the fault had given itself up and its surrender had gone unnoticed*!

Figure 3.40 shows some commercially available burn-down sets. Not all burn-downs are straightforward and many fail completely because of environmental effects or difficulties related to dielectric materials.

The location and environment of the fault path itself are critical. For instance, the fault may lie inside an undamaged cable and be starved of

DIAGNOSIS AND PRE-LOCATION 65

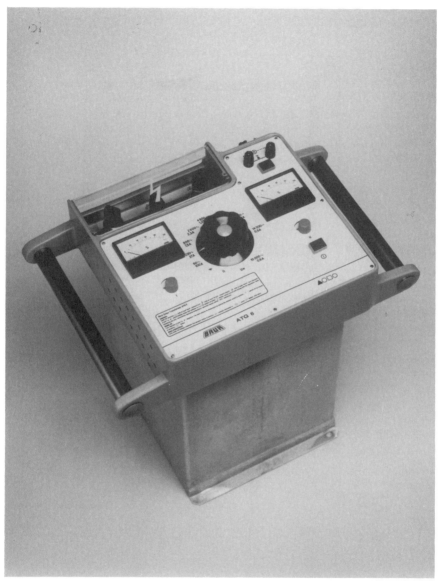

Fig. 3.40 (a)

oxygen. Such faults normally burn down quickly and well as the dielectric is easily carbonized and the carbon builds up easily in the confined space. Alternatively, the fault path may be a rough, broken area exposed to the air and therefore less easy to burn down.

66 UNDERGROUND CABLE FAULT LOCATION

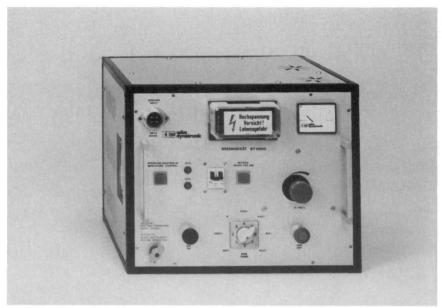

Fig. 3.40 (b)

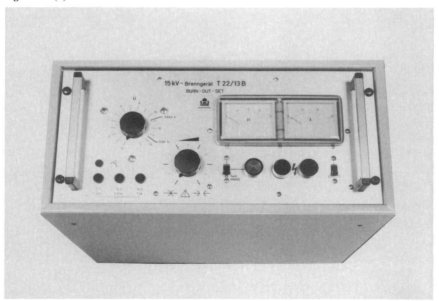

Fig. 3.40 (c)

Fig. 3.40 (a) The ATG 6 14 kV unit (*source:* Baur Prüf-und Messtechnik GmbH & Co. KG, Austria). (b) The BT5000 14 kV unit (*source:* Seba Dynatronic GmbH, Germany). (c) The T22/13B 15 kV unit (*source:* Hagenuk GmbH, Germany)

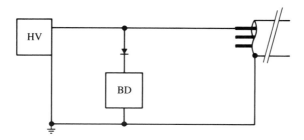

Fig. 3.41 Parallel operation of a high voltage test set and a burn-down set

A fault that can be difficult or impossible to burn down is one that lies inside a joint with a fluid dielectric such as oil which constantly flows in to re-seal it. One difficult situation can often be turned to advantage. It is very common for telecommunications and control cables to be faulty due to ingress of moisture, resulting in all conductors showing medium resistance readings, i.e. leg to leg, pair to pair and leg/pair to sheath or ground. The use of a burn-down set from the lower voltage group can dry out one or more of the conductors which can be used as healthy conductors for a bridge test or may show small differences on a PE/TDR trace. With regard to dielectric materials, paper burns down well and most of the plastics can be burnt down. However, polyethylene melts but cannot be burnt down to a conductive carbon path.

Finally, situations arise in which a high resistance cable fault cannot be broken down with a burn-down set. It is then necessary to start the burn-down process using a high voltage test set. This must be a set that does not trip out when breakdown occurs, but self-regulates and maintains a few milliamps through the fault. The unit can be left connected for a long time which can be several hours and may indeed achieve a successful burn-down on its own. However, it is more likely that it will be quickly discharged and disconnected and replaced by the burn-down set. If this 'picks up' the burn, so much the better. If not, it has to be accepted that the burn-down will not succeed and an alternative approach must be used.

There are some arrangements found in test vans and trailers where the HV test set is connected in parallel with a burn-down (BD) set that is protected by a rectifier, as shown in Fig. 3.41. This allows the heavier current available from the burn-down set to flow as soon as the high voltage breaks the fault down.

Instead of building up a carbon bridge which can be subsequently blown apart by surging, it is sometimes desirable to produce a very solid carbon bridge or, better still, a metal-to-metal weld, as shown in stage 4 of Fig. 3.39. This is when it is intended that the audio frequency methods of

68 UNDERGROUND CABLE FAULT LOCATION

Fig. 3.42 Promoting burn-down between two conductors

pin-pointing should be used. For this purpose a very low fault resistance is required, preferably between two conductors, so that the 'twist method' can be used (see Chapter 4).

A ploy that can be used to promote burn-down between two conductors when only one conductor is faulty to earth is to earth down one or more conductors at the test end so that, at the point of fault, stress is applied between conductors as well as between the conductor and earth (see Fig. 3.42).

When burning down to create a massive solid connection, the lower voltage a.c. outputs can be brought into play. The very heavy currents thus produced create a lot of heat at the fault, usually bringing about the desired result.

As a matter of interest, a steady burn using a.c. is indicative of the stable nature of the path being created because, if that path is at all fragile or liable to crack, the burn will often cease at a voltage zero, the applied alternating voltage then standing across the gap being too low to re-strike the arc.

3.5 Transient methods

In the sixties a method became available that made possible the pre-location of very high resistance and flashing faults which had hitherto been extremely difficult to find. This was concerned with the display and analysis of a voltage transient set-up in a cable when a high voltage capacitor was discharged into a fault. The source for this test is the normal surge generator used for pin-pointing faults by the acoustic method. Figure 3.43 shows a surge generator (SG) connected to the faulty core and a high voltage capacitor C commoned with the high voltage lead. The 'cold' end of this is connected to the PE/TDR oscilloscope via the low voltage capacitor c, which is usually incorporated within the set. C and c constitute a capacitor divider which delivers a safe low voltage signal of the same form as the high voltage phenomenon present on the cable core. It must be stressed that not all PE sets/TDRs were or are equipped with an input designed to receive externally generated voltage signals. This should always be borne in mind when choosing equipment.

DIAGNOSIS AND PRE-LOCATION   69

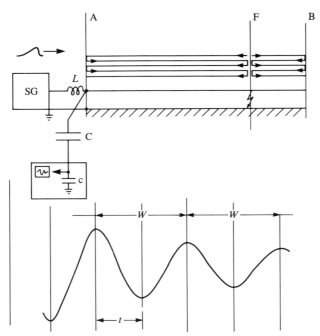

Fig. 3.43 Voltage transient from high voltage discharge into fault

Referring again to Fig. 3.43, the high voltage surge wave travels along the cable until it reaches the fault. After a certain time lapse the fault ionizes and arcs over and two voltage transients are set up which travel from the site of the fault in both directions, one towards the far end and one towards the test end. This latter is the transient which will be displayed on the oscilloscope. When it arrives at the source end, it is reflected with no change of polarity by the inductance L placed at the surge generator output for that purpose. On arriving back at the fault position, it is again reflected, this time with a change of polarity, by the fault arc itself which appears as a permanent short circuit because the duration of the arc is very long in comparison with the microseconds involved in the time of flight of the transient wave. This reflection and re-reflection continues until the energy in the transient has been dissipated due to losses in the cable.

As shown, the resulting trace displays constant wavelength w but diminishing amplitude. The wavelength is related to the fault distance which, in terms of time, is $1/2\,w$, i.e. the time taken for the transient to travel from F to A and back to F. Thus the distance to the fault is given by

$$D = t \times \frac{V}{2}$$

70 UNDERGROUND CABLE FAULT LOCATION

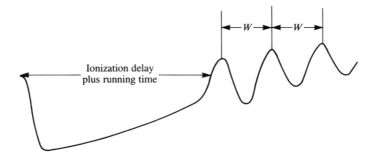

Fig. 3.44 Waveform of voltage transient likely to be found in practice

where t is the time of flight in microseconds and V is the velocity of propagation in metres or feet per microsecond.

In discussing and using transient methods of pre-location, the effect of the *ionization delay* must be fully understood, as this often gives rise to misreading and error. The basic fact is that, until the arc strikes, the explosion at the fault does not occur and the transients required for analysis are not generated.

While the gap is waiting to ionize and break down, which can take many microseconds or, occasionally, a few *milliseconds*, the surge wave is still travelling along the cable. The arc may strike when the surge wave has just passed the fault or has travelled to the end, been reflected and is on its way back! Indeed, it often happens that the surge wave breaks the fault down on the return journey because its voltage has been doubled due to the positive reflection coefficient at the end open circuit. The waveform shown in Fig. 3.43 is idealized. Figure 3.44 shows a display that is more likely to be found in practice.

An important thing to remember with this and all other methods based on transient analysis is to examine traces on a higher range setting on the monitor and not simply to use a range somewhat greater than the cable length, as would be the case with PE/TDR pre-location. Otherwise the operator may be viewing and trying to analyse the meaningless stretch of trace that occurs before the features related to the fault breakdown. A wrong 'pre-location' can often waste hours of valuable time, and, in any case, it reduces the operator's faith in the equipment.

In the early days of using this method great difficulty was experienced in reading off measurements from the screen as the trace only appeared every few seconds, depending on the cadence of the surge generator. It was normal to use a Polaroid camera on time exposure to bracket several 'shots'. This resulted in errors due to consecutive traces not necessarily

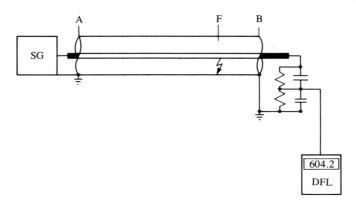

Fig. 3.45 Connections for digital fault locator

occurring in exactly the same position on the screen and, of course, the fact that the actual measurements had to be made directly from the Polaroid print.

Though these difficulties are worth mentioning, they are a thing of the past. All PE sets/TDRs/monitors now used for this purpose are equipped with memories so that a transient signal can be captured and analysed at leisure.

DIGITAL FAULT LOCATOR

In the late sixties, Dr P.F. Gale of the UK Electricity Council Research Centre undertook research into a method of capturing and analysing transients associated with cable faults, aimed at giving a direct readout of distance to fault in units of length. An instrument, the digital fault locator (DFL) was produced that could indeed give a direct readout of fault distance in the majority of HV power cable fault situations. It was widely used for some years in the United Kingdom, and to some extent in other countries, before being superseded by other developing transient methods such as impulse current (see page 73). Using the normal surge generator (SG) as source, the approach was to connect a fully compensated voltage divider at the *remote* end, as shown in Fig. 3.45, which fed a signal to the digital fault locator. This signal is a faithful reproduction of the voltage transient oscillating between F and B as previously discussed (see Fig. 3.43).

The oscillation is of the form shown in Fig. 3.46, where t_i is the ionization delay and t is the periodic time (equivalent to wavelength in previous discussions). The distance to the fault, d, is, as before, a half-wavelength multiplied by $V/2$, or

72 UNDERGROUND CABLE FAULT LOCATION

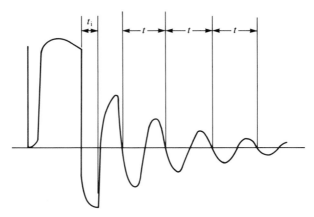

Fig. 3.46 Waveform of voltage transient

$$d = \frac{V \times t}{4}$$

An instrument came into being, therefore, that could measure the base frequency of the oscillating transient and display d in units of length. This, however, was not quickly or easily achieved.

The research which produced this method threw up several hitherto unknown facts related to transient measurements which were, of course, taken into account at the time, but which will undoubtedly have a bearing on future developments. Of these, the main ones were that both the velocity of propagation and the surge (characteristic) impedance were found to vary with frequency (see Fig. 3.47). This became evident when early field trials were carried out in which consistent measurements of good waveforms nonetheless produced errors of between 10 and 30 per cent.

This led to the incorporation in the instrument of a variable frequency oscillator with input voltage indication that is used to establish the velocity of propagation for the actual cable under test. With the remote end of the cable short circuited and open circuited in turn, the oscillator is tuned to find the harmonic, giving an instrument reading nearest to the apparent location. This is used to establish the correct velocity of propagation so that an accurate location can be made. The incorporation of a variable frequency source meant that the unit could also be used for pre-locating low resistance short circuits and open circuits by the standing wave method.

Although such faults are usually best found by PE/TDR techniques, this method is more effective and accurate than these on very long cables where a pulse suffers gross attenuation and dispersion. This is particularly true for telecommunications cables.

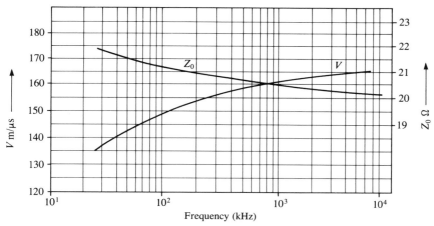

Fig. 3.47 Variation of the velocity of propagation and the surge (characteristic) impedance with frequency

In conclusion, then, while the digital fault locator had significant shortcomings, such as the necessity to operate at both ends of a cable and the need to calibrate the instrument, it enjoyed some degree of success. Some people had problems with it, but it also had many devotees who were finding (mainly flashing) faults that they could not easily find before.

The method has its place in the history of fault location not least because of the wealth of knowledge regarding the analysis of fault transients which was produced during its development. Bearing in mind the possibilities afforded by electronic technology of the nineties, it is more than probable that the last chapter on the direct digital readout of distance to fault is not yet written!

IMPULSE CURRENT METHOD

In the late seventies a development took place that has revolutionized power cable fault location. Dr P.F. Gale began considering the analysis of the *current transient* associated with the *voltage transient* which has been the basis of the transient methods described thus far. This approach, called the *impulse current method*, has many attractions, not the least of which is *safety*, the current signal being derived from a simple linear coupler normally attached to the 'earthy' return lead of the surge generator or HV test set as shown in Fig. 3.48. Note also that the linear coupler has an earthed screen between the secondary and the earth return (primary) because, in certain circumstances, the potential of the earth return can be raised.

This completely does away with the blocking capacitor/capacitor divider

74 UNDERGROUND CABLE FAULT LOCATION

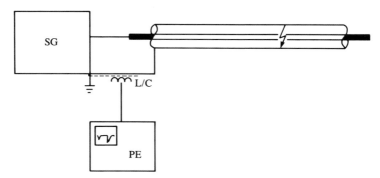

Fig. 3.48 Connections for the impulse current method

which has to be connected to the 'hot' lead. Another major advantage of looking at current displays is that, although the linear coupler is nothing more than a loop of wire run close to the primary lead for a few centimetres, the primary currents involved are large and, because the signal in the secondary is produced by a rapid rate of change of current in the primary, distinct 'spiky' trace features are produced. The effect is also enhanced by the high frequency response of the linear coupler which accentuates the leading edges of the trace features.

This compares with the less definite voltage waves discussed so far. Add to this the fact that, during the seventies, storage/memory oscilloscopes were available which made possible the proper analysis and measurement of transient phenomena and we have a powerful technique capable of pre-locating every type of power cable fault.

The philosophy of this approach is sound. In power cable fault location the instrument we *cannot do without* is a surge generator, normally used for pin-pointing. A 'black box' is added to this to make the set capable of pre-location as well.

When the method came into general use some manufacturers produced separate transient recorders to be associated with existing PE sets/TDRs and some produced dedicated impulse current equipment. Most instruments available nowadays have the necessary transient capture and measurement circuitry built in and are also capable of advanced PE/TDR operation. The linear coupler is sometimes built into the surge generator but is also available separately, as it sometimes has to be attached to other parts of the circuit, as will be seen later.

At this juncture it is proposed to cover the basics of transient behaviour in cables, related specifically to fault location by the impulse current method. The anatomy of the fault itself must first be considered.

A so-called 'flashing fault' behaves as a spark gap which will break down

DIAGNOSIS AND PRE-LOCATION 75

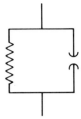

Fig. 3.49 Equivalent circuit of a fault

if a high enough voltage is applied to it. On the other hand, a short circuit fault of very low resistance cannot arc over. Most faults, however, fall into another category, i.e. they consist of a resistive path, sometimes of very high resistance. As practice indicates, these faults will pass current if stressed, but they will also arc over. The equivalent circuit of a fault is therefore as shown in Fig. 3.49, a spark gap in parallel with a resistance.

When considering *current* transients as opposed to *voltage* transients one has to take a different view of the trace features produced.

A positive voltage transient produces a negative current transient and the coefficients of reflection are of the opposite sense, as shown in Table 3.6.

Table 3.6 Features of voltage and current transients

Transient	Feature	Coefficient of reflection
Voltage	Open circuit	+ 1 (maintains same polarity on reflection, voltage doubling occurs)
Current	Open circuit	− 1 (inverts on reflection)
Voltage	Short circuit	− 1 (inverts on reflection)
Current	Short circuit	+ 1 (maintains same polarity on reflection, current doubling occurs)

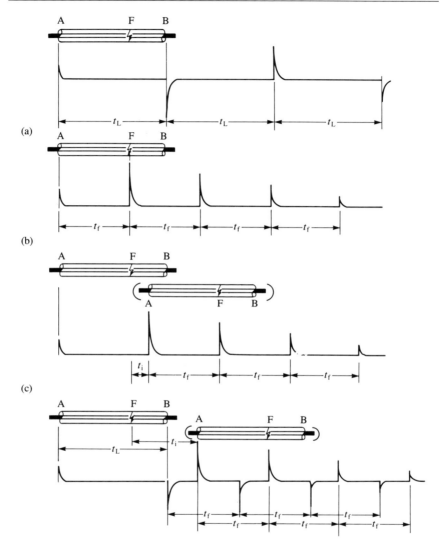

Fig. 3.50 The waveforms shown here are the traces produced by current transients in the same cable for the conditions (a) open circuit fault, (b) short circuit fault, (c) high resistance fault, (d) high resistance fault with long ionization delay

The waveforms shown in Fig. 3.50a to d are the traces produced by current transients in the same cable, length L, when a voltage surge is applied by a surge generator (which is effectively a short circuit to the transient).

Time lapse

Most engineers who are experienced in PE/TDR cable fault location normally have difficulty in changing their thought processes to properly consider and analyse transients. A basic consideration with PE/TDR fault location is that the maximum time lapse considered is the pulse flight time t_L from the test end of the cable to the extremity. All other times, i.e. to joints, tees or faults, are, by definition, less than t_L. Therefore the PE/TDR practitioner limits considerations to a zone within this time span t_L – any longer time lapse represents something 'beyond the end of the cable'.

When converting to transient analysis the operator must think entirely in terms of *time lapse* during which the transient is oscillating back and forth between the start point and some mismatch such as a fault or end. It is essential to consider a sequence of events as the transient travels along the cable.

Therefore trace (a) (open circuit) in Fig 3.50 is produced by current pulses resulting from a voltage transient starting at end A and travelling to end B, where it is reflected. It returns to end A and is re-reflected etc., its amplitude being progressively diminished by cable losses.

Note that, as we are dealing with *currents* and the start pulse at A is shown positive, the signal inverts on being reflected from B, (coefficient of -1 as stated above). This pattern continues in time, the signal being inverted each time it is reflected from B. There is no fault on this cable or, if there is a fault, the transient has not made it arc over; i.e. as far as the transient's performance is concerned, there is no fault.

Trace (b) (short circuit) shows the current signal from the linear coupler when the voltage wave is reflected and re-reflected from a zero ohm fault at F. Current doubling occurs, thus increasing the amplitude of the features. The end of the cable is not seen as the surge cannot travel beyond F.

Trace (c) is from a high resistance fault. Here, for the first time, the effect of ionization delay is noticed. Considering events *sequentially* (as emphasized earlier), it can be seen that the surge reaches the fault and passes on. The voltage wave has stressed the fault but not broken it down. At some point in time after this, but before the surge reaches the end B, the fault *does* break down. Now the two transients mentioned earlier are created, the one oscillating between F and B and not being seen on the oscilloscope, and the other oscillating between F and A producing the display depicted.

Note that no end feature is apparent. This is because the fault arced over, creating a short circuit and thus blocking the return path for the applied surge wave which is now trapped between F and B and cannot be seen as it is prevented from returning to A.

Finally, trace (d) results from a high resistance fault similar to that in (c) but, in this case, the ionization delay is longer. In fact, the breakdown occurs *after* the voltage wave has passed the fault position on its return journey. It is therefore allowed to reach A, giving rise to an end feature. Re-reflections of this now appear at intervals of t_f as it oscillates between end A and the fault. The transient generated by the breakdown produces short circuit features identical in form and spacing to those in trace (c) but shifted later in time.

Transient situations

Such is the influence of PE/TDR practice on engineers in the field that most operators are used to working with *negative-going* short circuit features and *positive-going* open circuit ones. The fact that current derived features are in the opposite sense can create confusion. Most manufacturers simply invert them so that they match PE/TDR convention. Hereafter, therefore, all illustrations will be treated the same way. Short circuit faults will show *down* and open circuit faults will show *up* (for the first reflection).

It will be apparent by now that several transients can *coexist* in a cable. We may not only be viewing the trace produced by the *injected* surge but also those set up by the fault arc. This is another factor that makes impulse current philosphy different from PE/TDR philosophy. In PE/TDR work the applied pulse is a low voltage one which does not break down a fault but is simply reflected from it and other mismatches. Therefore, except for its own re-reflections, this is the only pulse travelling in the cable.

Lattice diagrams are most useful in showing the progress of all the transients present in a cable, especially if aligned with the resulting traces. Figure 3.51a to e shows a series of lattice diagrams illustrating the same situations shown in Fig. 3.50a to d. They represent current transients but the traces shown are in conventional format, i.e. short, *down*; open, *up* (first reflection). The reflection coefficients are given at each mismatch.

In the open circuit situation in Fig. 3.51a there is only one pulse present, the injected one, which is launched from A and travels past the fault to the end B whence it is reflected back inverted to be viewed at A. It is again reflected but without inversion this time. This process continues until all the pulse energy is dissipated.

It will be noted that the display between the start pulse and the first reflection is the same as a PE/TDR display. The high voltage pulse producing this current pulse behaves in just the same way as the low voltage pulse from a PE set/TDR. As such, the trace can be trusted to obtain a correct measurement to the open circuit, providing the operator is absolutely sure that the fault has not broken down and produced other transients.

DIAGNOSIS AND PRE-LOCATION 79

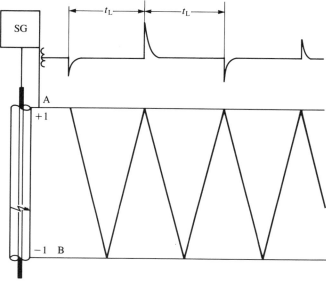

Fig. 3.51 (a)

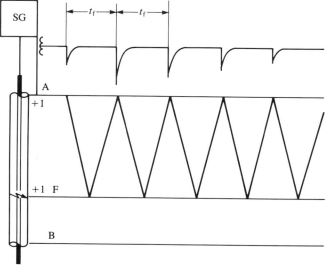

Fig. 3.51 (b)

In the short circuit situation in Fig. 3.51b the injected pulse is again the only one present, travelling from A to the short circuit fault at F. Here all its energy is reflected back towards A with the same polarity. No energy

80 UNDERGROUND CABLE FAULT LOCATION

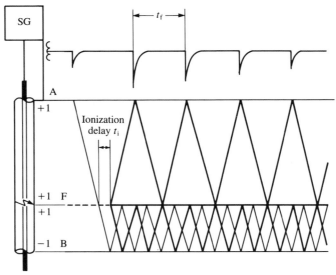

Fig. 3.51 (c)

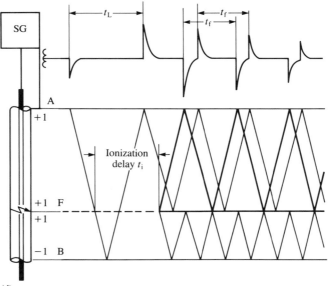

Fig. 3.51 (d)

can go on to B to show the end. The pulse bounces back and forth between A and F, giving diminishing trace features of the same polarity. As in the previous trace, the first part of the display is the same as a PE/TDR display and, for the same reasons, a correct measurement can be taken from it.

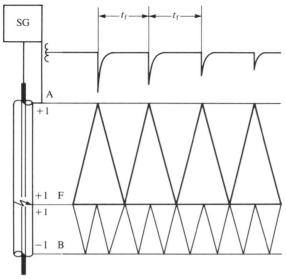

Fig. 3.51 (e)

Fig. 3.51 A series of lattice diagrams representing current transients with the associated traces shown in conventional format for the conditions (a) open circuit fault, (b) short circuit fault, (c) high resistance fault, (d) high resistance fault with long ionization delay (e) flashing fault 'relaxation' test

Once more it should be stressed that no breakdown must have taken place. If it has, the first part of the trace is not only meaningless for measurement but totally misleading.

In the high resistance fault situation in Fig. 3.51c a surge is launched into the cable at A which passes the fault but *does not break it down*. After a time lapse of t_1, the fault decides to break down. The two transients are now created which travel from the fault site to both ends of the cable, the one travelling to end A producing the trace shown. The other one going to end B is not seen and neither is the launched pulse, which finds its return journey blocked by the new short circuit, the fault arc, and is also forced to oscillate between F and B.

The trace is a true *impulse current* trace and does not relate at all to a PE/TDR trace. It is essential to search for and properly identify the *first* breakdown feature and measure from it.

In the high resistance fault situation in Fig. 3.51d with a longer ionization delay, as in the previous case, the launched surge passes the fault without breaking it down but this time there is a longer delay before it actually does – so much so that the surge passes the fault again before breakdown occurs after the ionization delay t_i. This results in two things happening, both very significant. Firstly, the surge is allowed to return to A so that the trace feature for end B is seen. Secondly, it is now trapped between F and A, producing the multiple reflections shown which maintain their sense and are separated by intervals of t_f, as are the negative-going features produced by the breakdown pulse.

There is a very important transient situation that has not yet been mentioned. This is where a test voltage is applied to an intermittent or 'flashing' fault which does not break down until a certain voltage level is reached. At this level the fault arcs over and the combined capacitance of the surge generator and the cable core is discharged violently into the fault, giving rise to the two familiar transients which travel each way from the fault to the terminations. *These are the only transients in the cable* as no surge wave has been injected.

Therefore the very first trace feature that appears is the breakdown pulse arriving at the test end. This is the only situation in impulse current testing where a measurement can be made *from the first feature* (whereas, in PE/TDR work, measurements are *always* made from the start of the trace).

This 'd.c.' or 'relaxation' test is most useful in practice because the impulse current set can be set up and armed, voltage applied and a result obtained from the one 'shot' taken as the fault breaks down. The trace produced is invariably clear, clean and unambiguous. Figure 3.51e is the lattice diagram for this situation and Fig. 3.52 shows the connections for the test.

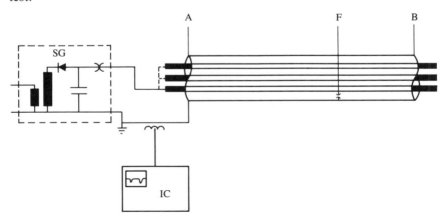

Fig. 3.52 Test connections for the 'relaxation' test

As the voltage is raised the fault gap is stressed but nothing happens visually before the fault breaks down. No surge has been applied and no transient exists. There is simply a d.c. potential standing on the faulty core.

Then, when the fault *does* arc over, all the energy pent up in the connected capacitance is released into the cable and the transients shown in Fig. 3.51e are launched, the one which travels towards end A producing the trace shown.

Of all the traces produced in impulse current fault location, this trace is the clearest, the most predictable and the easiest to analyse and measure from. There are no complications due to other transients as no surge has been applied, and ionization delay does not appear on the trace. The very first event is the breakdown pulse; therefore the operator can be sure that it is safe to measure from this first trace feature. Thus in one 'shot' a pre-location is made of a type of fault, the 'flashing' fault, which was almost impossible to pre-locate in the past.

PRACTICAL IMPULSE CURRENT FAULT LOCATION

Normal procedure

When performing a pre-location by this method it is usually helpful to carry out a few diagnostic checks using the impulse current equipment. This in no way invalidates the procedures laid out previously in Sec. 3.1 but it produces traces that will assist in interpreting the pre-location traces subsequently produced.

Figure 3.53 shows a surge generator and impulse current set connected to a three-core cable with a complex fault on it at F. Core 2 is healthy but cores 1 and 3 are faulty, core 1 having a high resistance fault on it and core 3 a flashing fault. It is assumed that these conditions have already been

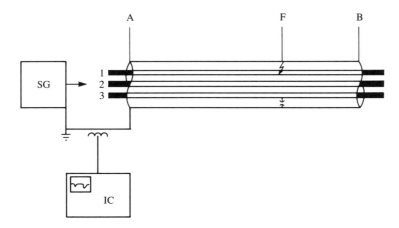

Fig. 3.53 A surge generator and impulse current set connected to a three-core cable with a complex fault

84 UNDERGROUND CABLE FAULT LOCATION

diagnosed earlier, but checks are being made as suggested above. With the surge generator connected to core 2, a low voltage of between a few hundred volts and, say, 2 kv is set, the impulse current set armed and a single surge applied. This triggers the set and a trace is produced like the one shown in Fig. 3.54.

As a relatively low voltage has been applied to a healthy core, no breakdown occurs. This being so, the trace produced is of a 'PE/TDR' type and the first part of the trace between A and B represents the total length of the cable. The other feature seen at R is a re-reflection of the one at B. Note that inversion takes place as these are *current* pulses.

Without causing too much inconvenience it is normally possible to verify this measurement and prove continuity at the same time by applying the same low voltage surge with a short at end B. Indeed, it is often better to do this for all three cores before the safety earth at end B is removed for testing.

This test produces a trace as shown in Fig. 3.55 with a large negative feature at B and a non-inverted re-reflection at R. Again, a 'PE/TDR' type of trace has been created as the surge has been fired into a known short circuit at B. The measurement from A to B in this case will be identical to the one in the open circuit case. This can easily be demonstrated by recovering the first trace from the instrument memory and superimposing one trace on the other, as shown in Fig. 3.56.

After making these preliminary checks, the operator is absolutely sure of the total length of the cable and that the core *and sheath* are continuous. (It should be remembered that PE/TDR is a two-pole test and an open circuit on the sheath will show as a positive trace feature.) The operator also has a good idea of how best to set up the impulse current set with respect to amplitude, sensitivity, expansion and range, etc. Attention can now be paid to carrying out a pre-location on cores 1 and/or 3.

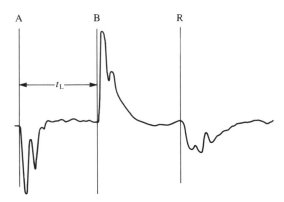

Fig. 3.54 'PE/TDR' type trace showing end of cable

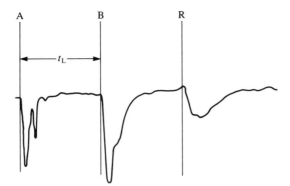

Fig. 3.55 'PE/TDR' type trace showing end shorted to earth

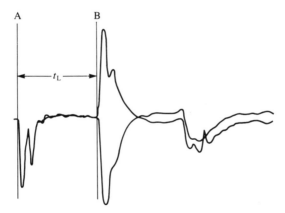

Fig. 3.56 Superimposing one trace upon another to accurately measure the cable length

Because such a clear and unambiguous trace can be produced by a flashing fault as explained earlier, it would be best to look at core 3 first. The surge generator should be set up to produce high voltage d.c. with the capacitance in circuit. The impulse current equipment should then be selected and the unit armed. At the diagnostic stage it will already have been ascertained at what voltage the fault breaks down. This helps in deciding the surge generator voltage and capacitance settings and the impulse current equipment parameters. The surge generator voltage is now raised slowly until breakdown occurs and the impulse current set triggers, producing a trace like the one shown in Fig. 3.57. This is a practical version of the trace in Fig. 3.51 e and the same explanations apply.

In contrast with the traces in Figs 3.54 and 3.55, this trace is a true impulse current trace which bears no resemblance to a PE/TDR trace. The

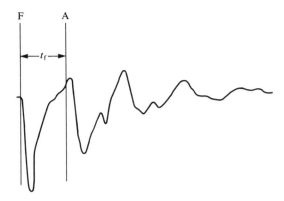

Fig. 3.57 Practical trace from 'relaxation' test on flashing fault

first significant feature is the fault breakdown and all other features occur later in time. The transient originates at the fault and we need it to travel to the test end A, go back to the fault position F and return once again to give us the second feature to measure to. This measurement is shown as t_f.

At this stage in the pre-location a calculation can be made of the distance to the fault. For example, say the cable length L is 2.2 km and the two measurements taken are

$$t_f = 20.62 \ \mu s \text{ and } t_1 = 27.5 \ \mu s$$

The distance to the fault is then

$$\frac{t_f}{t_1} \times L = \frac{20.62}{27.50} \times 2200 = 1649.6 \text{ m}$$

For the time being it is assumed that there is no connection cable of any significant length. If there is, its length in microseconds can be ascertained and this value subtracted from the measurements.

In Fig. 3.57, note the exact point to which the measurement t_f is made. It is taken to the start of a small positive 'bump' occurring just before the negative-going front of the large feature at A which is the arrival of the first reflection of the breakdown pulse at the test end. This 'bump' results from the time delay introduced by the inductance of the surge generator itself together with its connections and can therefore produce errors of measurement if not recognized.

The form of the trace features representing the reflections of the transient from the test end A is further explained in the later section on 'loop-on, loop-off'.

If an accurate route length is not available, the distance to the fault D_f is calculated as usual by

$$D_f = t_f \times \frac{V}{2}$$

In this case, if the value of $V/2$ for the cable is taken as 80 m/μs this becomes

$$D_f = 20.62 \times 80 = 1649.6 \text{ m}$$

The pre-location is therefore complete for core 3, but what of the high resistance fault on core 1? It could be said out of hand that this fault will be at the same location. However, although multiple faults usually lie at the same point, this can not be assumed. *Nothing* should be assumed in fault location, otherwise the logical build-up of evidence is destroyed.

The fault on core 1 should now be tackled. As it is a high resistance fault it will not sustain a high applied d.c. voltage, simply passing a steady leakage current. The surge generator is therefore set up for 'Surge' and a voltage of, say, 5 to 10 kV set. The impulse current set is armed and the surge generator started.

A relatively slow rate of surging, e.g. one shot every 5 to 10 seconds, should be selected so that, in between surges, the operator has time to rearm and perhaps adjust the surge generator voltage, impulse current set range and sensitivity settings as necessary to obtain the best possible trace. Alternatively, these initial adjustments can be made while applying surges with the 'single shot' button if such is incorporated in the surge generator. A good trace is one like the one in Fig. 3.58a.

If there seems to be no obvious breakdown feature, it may be either that the fault has not broken down and more voltage is required for it to do so or that the fault may have broken down and the ionization delay time is extremely long. To check this a very large range setting should be selected (see Fig. 3.59).

Figure 3.58b shows a trace produced when too small a sensitivity has been set. On the other hand, too high a sensitivity setting will produce a trace like that in Fig. 3.58c.

Figure 3.59 represents a very common situation which causes operators a considerable amount of trouble, particularly when inexperienced. In this case the ionization delay is so long that only the end feature and its reflection can be seen on the screen. The fault breakdown occurs 'off the end of the screen' in terms of time lapse. This long ionization delay can result from the fault just feeling contrary.

It can also result from there being too low an energy content in the applied surge (see Fig. 4.2).

88 UNDERGROUND CABLE FAULT LOCATION

Furthermore, as the actual breakdown pulse is not present on the screen, the operator, thinking that it must be there somewhere, can easily take a measurement from a promising looking negative feature and arrive at a totally wrong pre-location with the result that:

1 The operator's reputation is not enhanced.
2 The efficacy of the equipment is called into question.
3 People digging holes in the wrong places are not amused.

The significance should now be realized of the importance of adjusting voltages and settings in order to obtain a trace similar to that in Fig. 3.58a. Any potentially interesting trace can always be held in memory until a better one is captured.

Figure 3.60 shows a good example of a trace produced when the fault breakdown occurs after the surge has passed the fault position and has been reflected from the end B. Reference back to Fig. 3.51d will provide a reminder that there are two transients present in this situation:

- The transient produced by the fault breakdown which results in the *negative* trace features and

DIAGNOSIS AND PRE-LOCATION 89

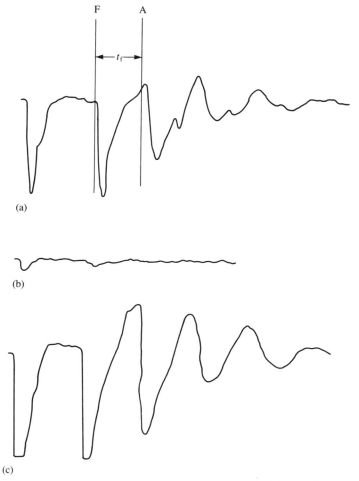

Fig. 3.58 Traces from a high resistance fault (a) sensitivity setting correct (b) sensitivity setting too low, (c) sensitivity setting too high

- The applied pulse which is producing the *positive* features having been trapped between source and fault (A and F), after the point in time at which the breakdown occurred

Of these, the former is by far the most prominent as it results from the very high current that flows when the fault arcs over, discharging the surge generator capacitance. The positive features resulting from the applied pulse reflections are weaker and often too small to recognize. The 'zone of interest', F to A, moves forward or back in time and, very often, the

90 UNDERGROUND CABLE FAULT LOCATION

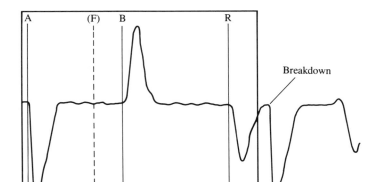

Fig. 3.59 Long ionization delay, fault feature 'off' the end of the screen

breakdown pulse at F is mixed up with the end feature at B. The operator should simply choose the clearest and most indicative trace, place it in memory and measure t_f as already described.

One very important point to remember is that, when comparing traces on the screen by superimposing one upon the other, *the breakdown pulse is the*

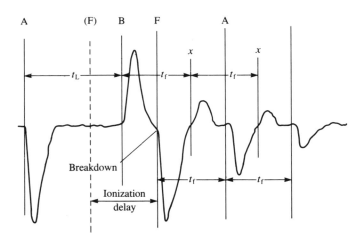

Fig. 3.60 Trace produced when fault breakdown occurs after the surge has passed the fault position and has been reflected from the end *B*

one key common feature. All features before this should be disregarded when coinciding the traces because, inevitably, the ionization delay will be different for shots taken at different times. Therefore it is the two *breakdown* features that should be brought together, *not* the two *start* features.

In the case under consideration the measurement of the fault distance for this high resistance fault on core 1 will be the same as, or similar to, the distance calculated for the flashing fault on core 3, confirming that both faults lie at the same point.

Difficulties with measurement—'loop-on, loop-off'

The large negative breakdown feature is usually clear and easy to measure *from*, i.e. the leading edge departs the x axis quite steeply. The first reflection, however, which it is required to measure *to*, can present difficulties in that the leading edge may turn down more gradually or go up before going down. These variations are produced by the inductance of equipment connected at the test end, that is to say, the surge generator itself and its connecting leads or cables. Different surge generators will present slightly different delays, but greater effects are produced by the test cable.

It follows then that, if a measurement is taken from the point at which the first reflection turns down (shown as * in Fig. 3.61), there will be an error due to these delays. Of course, the delay for any particular test rig can easily be measured in a test situation and the resulting time delay subtracted from all measurements made when using that rig. Figure 3.61 shows this effect.

As the impulse current method was becoming more widely used in the late seventies and early eighties, many test rigs were in evidence, varying

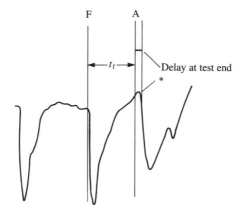

Fig. 3.61 Error produced by delay at test end

92 UNDERGROUND CABLE FAULT LOCATION

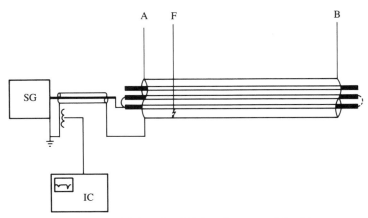

Fig. 3.62 Test connections with length of high voltage coaxial cable

from a surge generator connected directly on to the cable under test to equipment mounted in a tailor-made test van with built-in cable drums. To take out the effect of the test end delay, a method was devised of 'marking' the cable termination at the test end or the remote end. This method is known as 'loop-on, loop-off' and can be applied in several ways. The procedure relies on applying a surge with the faulty cable in a certain circuit configuration and memorizing the trace. Then the *circuit configuration* is changed and a second surge applied. The two traces are coincided at the breakdown feature and the point of divergence of the two traces marks the place in the circuit where the connection was made to change the circuit configuration.

In Fig. 3.62, the surge generator is shown connected to the faulty cable by a length of high voltage coaxial cable, as is usually the case in practice. With this arrangement it is only possible to 'mark' the *far end* of the cable.

The first circuit configuration is created by making a connection at A between the faulty core and a healthy core (or one that would break down at a higher voltage than would the designated faulty core). A surge is then applied and the trace memorized. The example shown in Fig. 3.63 represents a fault quite close to end A, a situation in which this *far end loop* method is particularly useful because, as can be seen, the features for the breakdown and its re-reflections are cramped up near to the breakdown feature, making measurement extremely difficult.

After switching off, discharging and earthing, a jumper is then connected between the faulty core and the same healthy core at the *far end*. Another surge is applied. It is important that the surge generator be set at the same voltage as it was for the first surge. This ensures that the trace features will be of similar magnitude in both cases and that the ionization delays will be similar. The second trace is shown in Fig. 3.64.

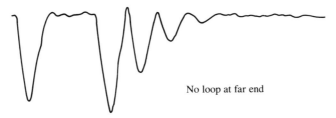

No loop at far end

Fig. 3.63 Trace with far end 'loop-off'

For the *first* shot which gives rise to the trace shown in Fig. 3.63, the current pulse created when the fault breaks down has one route back to the linear coupler at end A. However, when the *second* surge is applied with the loop on at B, the two pulses generated which travel away from the fault in each direction have routes back to the linear coupler. One, as before, travels directly along the faulty core back to end A, but the other one travelling towards end B can carry on to end A via the second, healthy, core. It takes longer to do so because it has further to travel, this difference being the distance between F and B, i.e. the *fault distance measured from end B*.

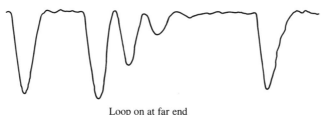

Loop on at far end

Fig. 3.64 Trace with far end 'loop-on'

The two traces are then superimposed as shown in Fig. 3.65, care being taken as usual to coincide the *breakdown features*. From this it clearly can be seen that the first breakdown feature (no loop) arrives first and the second breakdown feature (with loop) arrives later after a time lag of t_f as shown.

It is common practice also to use a *loop-on, loop-off* technique to 'mark' the near end A. However, if only one test lead is in use, the comparison of a trace resulting from a surge into just the faulty core and a trace resulting from a surge into a faulty and a healthy core (i.e. with a loop at end A) is not easy. This is because the applied surge will encounter very different impedances on the first and second shots and, also, on the second shot, its energy will be split so that it will probably take longer to ionize and break down the fault, making the alignment of traces more difficult.

The normal approach is to use *two* HV coaxial test cables approximately

94 UNDERGROUND CABLE FAULT LOCATION

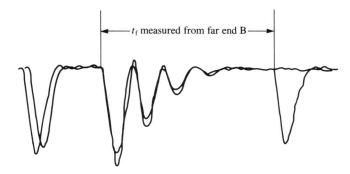

Fig. 3.65 The two traces are superimposed and the breakdown features carefully coincided

50 m long, as shown in Fig. 3.66. They are commoned safely at the surge generator connection by a 'splitter box' and connected to one faulty core and one healthy/relatively healthy core of the cable under test. With this circuit configuration, the problems encountered using one test cable only are overcome and the resulting traces are very similar up to the trace separation occurring when a loop is connected at A.

To carry out the *near end loop test*, a surge is applied with no loop at A and a trace such as that in Fig. 3.67 obtained.

A loop is then made between the faulty and healthy cores and a trace produced like the one in Fig. 3.68.

It will be seen that the trace is significantly different at the feature

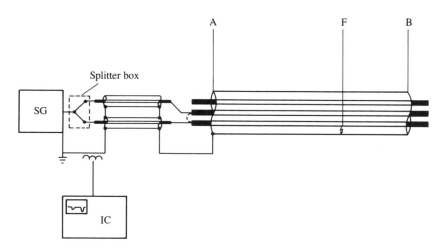

Fig. 3.66 Test connections for near end loop test using two coaxial connection cables

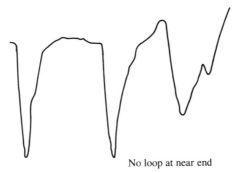
No loop at near end

Fig. 3.67 Trace with near end 'loop–off'

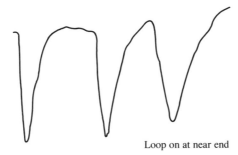
Loop on at near end

Fig. 3.68 Trace with near end 'loop–on'

representing the first reflection of the breakdown. This is due to the reflection factor at the junction of the test cable and the cable under test being changed by the application of the loop.

To measure accurately to this point of difference, the first trace (loop-off) should be recovered from memory and the two primary breakdown features coincided, as shown in Fig. 3.69. Note again that the start features do not line up due to the ionization time being different for each trace. As already stated, the important features to line up are the two breakdown features. All before this is irrelevant.

The difference between the traces will be significant and easy to recognize providing that the test cable is reasonably long in terms of time delay. A normal test cable about 50 m long represents a delay of almost half a microsecond, which gives enough space to the difference on the screen for it to be clearly recognizable as the difference just created.

The fault distance t_f can now be measured and it should be clear by referring back to Fig. 3.61 that no allowances need be made for time delays due to test cable/stray inductances.

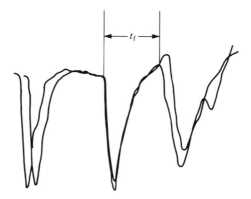

Fig. 3.69 The 'loop–on' and 'loop–off' traces can be superimposed

Summary

It can be said in theory, and in most cases in practice, that the *impulse current method* is capable of pre-locating all types of fault, normally from one end. This is not to say that there are not occasions when a test from the far end is useful or, indeed, necessary. In these cases, as has been shown, there is no need to transport all the test equipment to the remote end. A far end loop can be used.

Apart from situations in which the near end tests are inconclusive, it can also be a good idea to apply both near and far end loop tests if the very best accuracy is required for the pre-location. This can be the case when the suspect area is difficult to work in or gain access to, for example, sites on or near motorways, railways, rivers, rubbish tips, dense forest, runways.

The only limit to the application of the impulse current method would seem to be in testing those faults that have a higher voltage threshold than the surge generator in use, usually in the region of 30 kV. In fact, this is not the case as there are special test circuit configurations that can be used with the higher voltages derived from EHV (extra high voltage) pressure test sets on faults with breakdown voltages of the order of 80 kV. These tests are covered in Sec. 9.2. Also covered in this chapter are further impulse current tests for fault location on teed MV and HV systems.

ARC REFLECTION

General

Arc reflection is a seductive technique the basic philosophy of which appeals greatly to all fault location engineers who are, after all, only human and like the idea of *doing something* to the fault and immediately seeing and analysing the results. The whole approach is basically PE/TDR associated with fault treatment in that the operator views a trace without a fault and then makes the fault arc over which produces a visible difference on the screen.

The technique requires an HV source (which can be a burn-down set or surge generator), a filter unit and a suitable PE set/TDR or monitor. The source is connected to the cable under test via some form of filter/blocker which, while allowing the test voltage to be applied to the cable, admits the train of LV pulses from a PE set/TDR into the cable. The filter protects the PE set/TDR from the high voltage so that the test pulses are effectively launched into a live cable and their reflections received back.

Historical

In the mid seventies a system was developed, as shown in Fig. 3.70, whereby a normal burn-down set was connected to the cable under test via a filter with a signal connection to a PE set/TDR. Before the burn-down set is put into operation the PE/TDR trace represents a healthy cable (as far as the PE set/TDR is concerned). When the voltage from the burn-down set produces an arc at the fault, the trace will exhibit a negative feature at the fault, as shown in Fig. 3.71, *if the arc path has a low enough resistance.*

The fact that the PE set/TDR is being used in conjunction with a high voltage source does not alter what it can or cannot see. As explained in Sect. 3.4 on PE/TDR, a PE set/TDR will only show a reflection from a shunt path less than about ten times the characteristic impedance of the cable – a few hundred ohms at best.

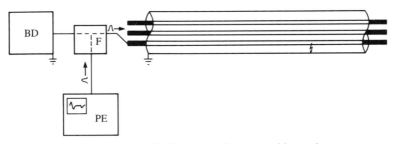

Fig. 3.70 Connections for arc reflection test using normal burn-down

98 UNDERGROUND CABLE FAULT LOCATION

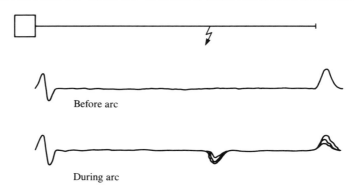

Fig. 3.71 Traces before and during arc

To produce a low enough arc resistance for a successful arc reflection test, an arc current of at least 3 A is required. This in turn means that a burn-down must have reached an advanced stage for such a current to be flowing. Therefore, in most such cases, a successful burn-down and pre-location may well have been possible anyway.

The other main problem with this method was that the burn-down set rating, and that of the filter, limited the maximum voltage which could be applied. The voltage specification of burn-down sets in normal use was, and is, in the region of 14 to 15 kV.

This approach, then, had only a limited success and was overtaken by further developments around 1980. These had to do with the incorporation of a capacitor working with the burn-down set, as shown in Fig. 3.72. (Though it was to come later, the direct step to the use of a surge generator was not made at this time because, even though the currents produced are well in excess of the current needed to produce a very low resistance arc, they flow for too short a time to be visible on a PE/TDR instrument screen in real time. Also, the voltage withstand of the filter would have had to be much higher.)

Figure 3.72 shows a conventional burn-down set with a capacitor connected across it and a controllable spark gap switching the output to the cable under test. This arrangement is basically a small surge generator producing a rapid succession of surges with low energy levels of the order of 20 J. The currents, however, are significantly higher than 5 A so they produce visible differences on the trace. Also, a burn-down set has a much larger power rating than a surge generator and this enables it to sustain repeated rapid surges.

The surge rate is varied to produce the best effect on the screen and will normally be in the region of 8 to 12 surges per second. This means that,

DIAGNOSIS AND PRE-LOCATION 99

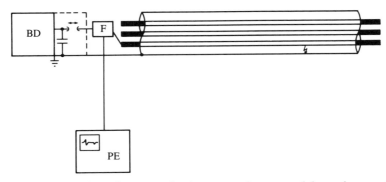

Fig. 3.72 Connections for arc reflection test using normal burn-down set with capacitor and controllable spark gap

because of persistence of vision and the phosphor of the screen, the operator sees an almost steady display showing, in effect, the healthy and faulty states simultaneously, as in Fig. 3.73. There is, of course, a flickering and wobbling as the depth of the negative trace feature varies with the changing fault arc resistance. Nevertheless, any visible difference produced by the application of the surges is enough to show where the fault lies.

The technique had reached a stage where it was enjoying a degree of success but still suffered from the following disadvantages:

1 The maximum applied voltage was still limited by the specifications of the burn-down set and the filter.
2 A good trace effect was not always easy to produce.
3 It was still necessary to possess a burn-down set.

This latter point was becoming more and more significant in the early eighties with the increasing popularity and use of the impulse current method which only requires a surge generator, the impetus of technical and commercial developments at that time being to do away with the heavy and costly burn-down set.

In the mid eighties, logically following this trend, came the emergence of systems using a surge generator in conjunction with a better protected filter

Fig. 3.73 Healthy and faulty states shown simultaneously

100 UNDERGROUND CABLE FAULT LOCATION

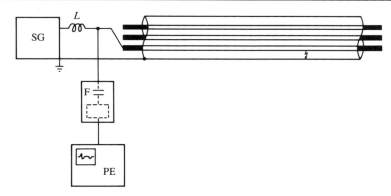

Fig. 3.74 Connections for arc reflection test using a surge generator with inductance and filter

capable of withstanding voltages as high as 50 kV. Such an arrangement is shown in Fig. 3.74. The significant addition here is the inductance L, the purpose of which is to prolong the duration of the arc so that it can be recognized on the screen in real time. Thus, at every surge, a negative blip appears at the fault position on the trace.

A further refinement has been added since in that the use of PE sets/ TDRs/monitors with transient recording facilities and multiple memories enables the trace representing the faulty condition to be captured in memory and superimposed on the one for the healthy condition, making accurate measurement extremely easy. This is shown in Fig. 3.75.

The main difficulty encountered in producing a good display is making sure that the pulses from the PE set/TDR are synchronized with the striking of the arc. Modern electronic techniques now make this possible. A manual delay feature can also be incorporated so that the operator has some degree of control in difficult situations.

Another approach to prolonging the arc is to incorporate a separate d.c. source in the filter unit suitably protected from the surge generator high

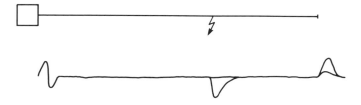

Fig. 3.75 PE/TDR sets/TDRs/monitors with transient recording facilities and multiple memories enable traces to be superimposed making accurate measurement easy

voltage. When an arc is struck, this arc stabilization unit is switched in so that an enhanced arc current of about 3 A is maintained for a few seconds. This is long enough for the 'difference feature' on the trace to be recognized. It is normal then to key the trace into memory so that it can be compared to the 'healthy state trace' and for a measurement to be made at leisure.

At the time of writing, in the early nineties, there exist several forms of arc reflection system as follows:

Burn-down set with capacitor and switch	+ filter	+ PE set/TDR/monitor
Burn-down set with capacitor and switch	+ filter	+ PE set/TDR/monitor with memory
Coupled HV and LV burn-down sets with capacitor and switch	+ filter	+ PE set/TDR/monitor with memory
Surge generator with inductance	+ filter	+ PE set/TDR/monitor
Surge generator with inductance	+ filter	+ PE set/TDR/monitor with memory
Surge generator with resistance	+ filter	+ PE set/TDR/monitor with memory
Surge generator	+ filter/arc stabilization unit	+ PE set/TDR/monitor with memory

Arc reflection testing

Carrying out an arc reflection test in the field need not be too difficult, providing the fault arcs over and produces some trace feature or difference on the CRO screen. Any difficulties lie in creating a good enough arc and in recognizing the trace feature. The actual test is a straightforward PE/TDR measurement.

In normal PE/TDR tests the only doubts concern identifying the feature to measure to. The measurement itself is easy. Therefore the techniques required in arc reflection testing relate more to the setting up, adjustment and control of the equipment than to the measurement. These techniques can be split up roughly into three groups:

- Real time tests
- Real time tests with memory facilities
- Tests with transient recordings

Real time tests

The equipment should be connected in accordance with the manufacturer's instructions and switched on, no voltage as yet being applied to the cable. Following the principles mentioned previously in Secs 3.1 and 3.2 on 'diagnosis' and 'pre-location', the PE set/TDR will already have been used to check the trace from the 'healthy' cable core and all matters such as length, continuity, propagation velocity, amplitude and balance settings, etc., will have been resolved.

Prior to the start of the arc reflection test, the PE set/TDR will be set up correctly and showing a trace. Also, the operator will be happy with the aspect of the trace regarding the 'end feature' and possibly other features representing joints or tees. All the aforementioned settings will be correct.

When the high voltage is applied to the fault, the operator will be trying to identify and 'mark' the point at which the 'fault feature' appears. The best way of doing this is to use a cursor. This will normally be a movable vertical line and the cursor shift control will be rotary or press button. Whichever is the case, the intention is to move the cursor as quickly as possible once the voltage is applied. To this end, therefore, the cursor should be set to the middle of the trace so that it has the best chance of being close to where the fault feature may appear.

When things are thus prepared, the high voltage can be applied. If the HV source is a modified burn-down set, a range should be selected that has a higher voltage than the withstand limit of the fault or, if no voltage threshold has yet been established, the highest voltage range should be selected. The voltage should then be raised slowly until a burn starts. Meanwhile, a careful watch should be kept on the CRO screen for the appearance of a 'blip' caused by the arc. It may be necessary to adjust the spark gap control as well. In any case, the manufacturer's instructions should be followed regarding burn-down control until a trace feature has been created and marked with the cursor. The fault distance is then measured in the normal way.

If the source is a surge generator, the PE set/TDR should be set up as just described. Then the surge generator should be started and its voltage raised slowly until breakdown occurs. The voltage should then be increased a little more so that a breakdown ought to occur with each surge. Again, the operator is trying to recognize and mark the position of any blip occurring on the trace. Unlike the feature produced by a burn-down set, this blip will only be visible for a fraction of a second, but the timing of its appearance is more predictable, i.e. it can only appear when a surge is applied.

Once the cursor has been positioned on the blip/fault feature, it only remains to measure the fault distance.

Real time tests with memory facilities

These tests are essentially the same as those just described except that the PE set/TDR/monitor used is capable of storing traces for subsequent comparison and analysis. This means that, as soon as a 'fault blip' appears, the trace can be stored. Success is then assured as the two traces for the healthy and faulty states can be recalled and superimposed and measurements made at leisure.

However, being able to store a trace when a sudden change occurs presupposes that the 'fault blip' remains visible long enough for the operator to be able to recognize it and hit a store button before it disappears. Of necessity, then, it is clear that this approach can only be used with equipment combinations that have the capability of prolonging the arc for several seconds. This means those rigs wherein the source is a modified burn-down set or a surge generator with an arc stabilization unit. The combinations with a surge generator plus inductance as source do not lend themselves to 'manual storage' as described above.

Tests with transient recorders

This final category represents outfits with either burn-down sets or surge generators as source but which are equipped to synchronize the PE/TDR pulses with the arc so that the PE set/TDR/monitor triggers and a display showing both the healthy and faulty state traces is frozen for easy analysis. Though this would appear to be the most effective method, it is only so if it consistently produces clear, unambiguous recordings on most tests without much operator involvement. This is quite an onerous requirement, but it does now seem that current electronic technology is making it possible.

The traces shown in Figs 3.71, 3.73 and 3.75 are indicative but somewhat idealized. Figures 3.76 and 3.77 are more typical of those encountered in the field.

Figure 3.76 is typical of a relatively straightforward situation on a cable less than 1 km in length where a good arc has been struck and the trace keyed into memory. Although there are minor trace divergences before the fault position, as usual, the main divergence marking the point of fault is easy to recognize and measure to.

The trace in Fig. 3.77 represents a flashing fault at a remote tee joint on a cable almost 2 km long. This is typical of a transient situation where the PE set/TDR/monitor triggers satisfactorily while surges of several bilovolts are being applied. It can be seen that, though the healthy and faulty traces

104 UNDERGROUND CABLE FAULT LOCATION

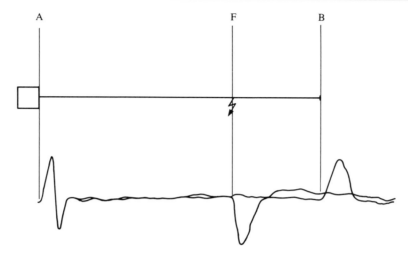

Fig. 3.76 Practical trace keyed into memory when arc has been struck

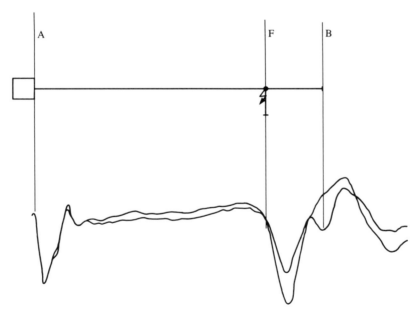

Fig. 3.77 Practical traces from a flashing fault at a remote tee joint on a cable almost 2 km long

diverge significantly at many points along the cable, the divergence at the point of fault, F, is quite clear.

Figure 3.78a to e shows some modern arc reflection rigs. Unless the arc reflection set is a dedicated module or unit, it is normally composed of one of several burn-down sets or surge generators working in conjunction with one of several PE sets/TDRs/monitors and the appropriate coupling unit. The list of equipment combinations given after Fig. 3.75 gives some idea of the many arrangements in use and the manufacturers' literature should be consulted carefully to assess their relative capabilities. Some examples of actual fault locations can be found in the Appendix to Chapter 9. The Appendix to this chapter is a discourse on the behaviour of transients in cables.

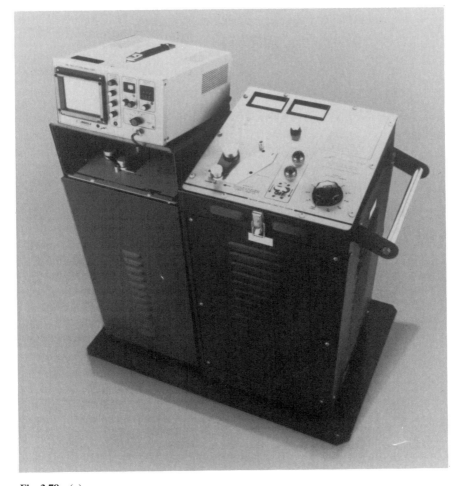

Fig. 3.78 (a)

Fig. 3.78 (b)

Appendix 3A Voltage and current waves in cables

It is frequently difficult for a fault finder to puzzle out the significance of an unusual oscillogram unless a very clear picture has been obtained of the behaviour of voltage and current waves in the cable system. Mathematical treatments are not always helpful because they can sometimes obscure the physical reality of the phenomena. An alternative approach is to consider the total energy in the cable as a whole.

Consider a length of cable, say single cored and coaxial, charged from a d.c. source to a voltage V. Disconnect the cable from the source and apply a short circuit to one end. Because the cable has length it cannot discharge instantaneously, a time at least equal to the length of the cable divided by the speed of light is required for the information of the short circuit to reach the remote end. The short circuit condition therefore travels along the cable at a fixed velocity representative of the type of cable. In fact, textbooks tell us that if the time for light to travel the length of the cable is $t(1)$ and if the cable has a dielectric constant (or relative permittivity) ε_r, then the time required for the information to reach the far end is $t(1)(\varepsilon_r)$. Call this time $t(c)$.

DIAGNOSIS AND PRE-LOCATION 107

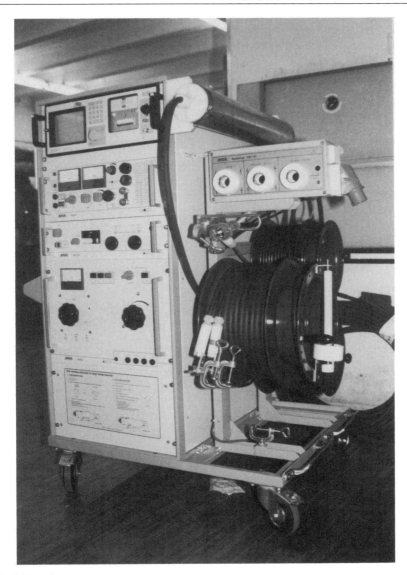

Fig. 3.78 (c)

From the instant of applying the short circuit to time $t(c)$ current will flow out of the cable into the short circuit. How much current? Initially the cable was charged to voltage V and if the capacitance of the cable is C, the energy initially stored in the cable is $0.5CV^2$ and the charge is CV. The current is then $CV/t(c)$ because charge CV = charge It. To evaluate I we still need to know $t(c)$.

108 UNDERGROUND CABLE FAULT LOCATION

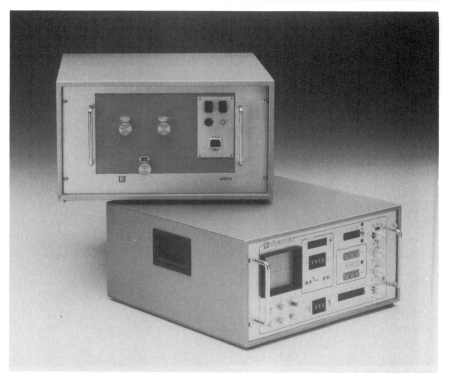

Fig. 3.78 (d)

However, let us first look at the end of the discharge period. Current I is flowing along the whole length of the cable. No mention has yet been made of the resistance of the cable and for the moment let us assume this resistance to be negligible. By definition the resistance of the short circuit is zero. Associated with the current I in the cable is a magnetic field which is a store of energy.

We are ignoring losses and so we must assume that the original electrostatic energy $0.5CV^2$ is converted to magnetic energy $0.5LI^2$, where L is the inductance of the total length of the cable. Therefore, we may write

$$0.5CV^2 = 0.5LI^2$$

or, more simply,

$$CV^2 = LI^2$$

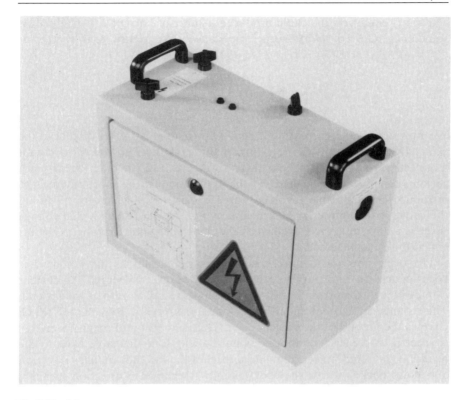

Fig. 3.78 (e)

Fig. 3.78 (a) The PFL–3000 Arc Reflection System (*source:* AVO Biddle Instruments, USA). (b) The LSG 32 associated with the CAF eta, in a test van (*source:* Seba Dynatronic GmbH, Germany). (c) The assembly incorporating the IRG 300 (*source:* Baur Prüf-und Messtechnik GmbH & Co. K.G., Austria). (d) The ARB 15 with the DIGICABLE (*source:* Balteau S.A., Belgium). (e) The M219 A.R.M. power separation filter (*source:* Hagenuk GmbH, Germany)

Hence,
$$I^2 = \frac{CV^2}{L}$$
or
$$I = V\sqrt{\frac{C}{L}}$$

If we rewrite $V\sqrt{C/L}$ as $V\sqrt{L/C}$, $\sqrt{L/C}$ is equivalent to a resistance or an impedance, which is in turn characteristic of the particular cable; that is
$$\text{Characteristic impedance } Z_0 = \sqrt{\frac{L}{C}}$$

110 UNDERGROUND CABLE FAULT LOCATION

The short-circuit on the cable produced a surge of current $I = V/Z_0$; Z_0 is sometimes referred to as 'surge impedance'. We return to $CV = It$ or $CV = (V/Z_0)t$ or $CV = Vt/\sqrt{L/C}$, so that

$$t = \left(\frac{CV}{V}\right)\sqrt{\frac{L}{C}} = \sqrt{LC}$$

We now see that a current V/Z_0 flows for a time $\sqrt{LC}$ to discharge the cable. During this period, $t(c)$, part of the cable near to the short circuit manifests zero voltage and part near to the far end (or open circuit end) sustains full voltage. The effect of the application of the short circuit is to introduce into the cable a voltage surge equal to $-V$ which travels along the cable, reaching the far end in time $\sqrt{LC}$. Associated with this voltage surge is a current surge travelling at the same speed and arriving at the far end at the same instant.

The cable appears to have been discharged inasfar as the voltage has been reduced to zero but the cable is full of magnetic energy. The current cannot flow into an open circuit so it attempts to fall to zero at the far end. This change in current in an inductance produces a back e.m.f. which charges the far end of the cable to $-V$. This conversion of magnetic energy to electrostatic energy continues along the length of the cable back to the near end. In detail, the current surge reaches the far end, changes direction and flows back towards the short circuit. A change in direction means a change of sign, that is to say a current surge of $-I$ flows from the open circuit end to the short circuit end. The direction of current remains the same in the external short circuit but $-I$ flowing from the far end has its associated voltage surge of $-V$ flowing with it and charging the cable to $-V$. When these two surges of $-V$ and $-I$ reach the short circuit the current falls to zero, leaving the cable sustaining a voltage of $-V$.

This is not a stable condition. We cannot have a short circuit and a voltage of $-V$ so a voltage surge enters the cable to eliminate the voltage of $-V$, i.e. the surge is $+V$. This voltage surge and its associated current surge $+I$ travel to the end of the cable and $-V$ and $-I$ return from the far end to give the conditions, when they reach the short circuit, of the current being zero and the voltage on the cable being $+V$, the conditions at the outset.

Alternate charging of the cable from $+V$ to $-V$ and back to $+V$ continues indefinitely in the absence of losses. In the real cable, however, the losses cause the magnitude of V to fall exponentially towards zero.

Note that we have seen a voltage surge of $-V$ returning (reflected) from the open-circuited end without change of sign but $+I$ reflected from an open-circuited end with sign reversal and a surge I is reflected from a short circuit with unchanged sign.

Another important point to note is that when the short circuit is applied to the cable the current flowing into the short circuit is V/Z_0. The cable then exhibits a source resistance of Z_0.

What happens if a resistance of Z_0 is put in series with the short circuit? The voltage V to which the cable is charged drives a current into the short circuit through the series Z_0 from its own source resistance of Z_0. The current into the short circuit is therefore $V/2Z_0$ so that a voltage surge of $-V/2$ and a current surge of $-V/2Z_0$ enter the cable. When they reach the far end the voltage on the cable is reduced to $V/2$. The surges are reflected and reduce the cable voltage by a further $V/2$ to zero. When the surges reach the near end, the current falls to zero leaving the cable uncharged. The cable has then been satisfactorily discharged in the time it takes for a surge to travel twice the length of the cable.

It is also clear that both voltage and current surges enter a termination of resistance value Z_0 without reflection. More generally:

1 A voltage surge entering a termination of value $> Z_0$ is partially reflected without change of sign.
2 A voltage surge entering a termination of value $< Z_0$ is partially reflected with change of sign.
3 A current surge entering a termination of value $> Z_0$ is partially reflected with change of sign.
4 A current surge entering a termination of value $< Z_0$ is partially reflected without change of sign.

4

Pin-pointing

Pin-pointing has been defined in Chapter 2 as a test that confirms the exact position of the fault overground and the two main methods have been listed as:

- *Shock wave discharge*, using a surge generator together with acoustic listening apparatus, and
- *Audio frequency*, with instruments for injecting and receiving a tone

While these two methods account for most fault confirmations, we must not lose sight of the basics of pin-pointing. This means recognizing and taking into account *any manifestation* of the presence of a fault.

Effects that can be noted at the site of a fault include:

- Change of voltage
- Change of polarity
- Change of current
- Modification of magnetic field
- Emission of sound/ultra-sound
- Temperature rise
- Chemical changes
- Olfactory changes
- Visual changes
- Partial discharge
- Emission of electromagnetic signals (HF, RF)
- Physical movement
- Microphony

Fault location is difficult and every possible means of pin-pointing should be looked at very carefully.

In this chapter, the headings under which pin-pointing will be covered are:

- Acoustic
- Electromagnetic
- Pool of potential
- Magnetic field
- Audio frequency

Other methods such as *thermal* and *partial discharge* will be covered in Part Three, 'specialized areas'.

4.1 Acoustic

In power cable fault location the vast majority of pin-pointing tests are carried out using a surge generator to create noise and vibration at the fault site. The surge generator used for this is basically a variable HV d.c. source connected to a high voltage capacitor bank, the output being via some sort of spark gap or triggered contactor.

In the early days of fault location the transformer, rectifier, capacitor and gap were discrete components mounted up and connected on site. The gap was not triggered. The contact separation was simply adjustable, and the length of the gap determined the rate and voltage at which it discharged. Figure 4.1 shows the basic construction of a modern surge generator.

Now all commercially available sets are enclosed for safety and a 'dump' switch (D) is incorporated which discharges the output to earth via a resistance (R) when the instrument is switched off. The rating of the transformer must be greater than that of a straightforward HV test set as it must have enough output capacity to charge the capacitor within a few seconds.

The surge generator works by charging up the capacitor C with the switch S open. This switch or contactor has hardened contacts for long life and is solenoid operated so that it can be closed and opened by a timing circuit T which enables the operator to vary the cadence of the switching to give a closure once every 1 to 10 seconds (approximately). The value of

114 UNDERGROUND CABLE FAULT LOCATION

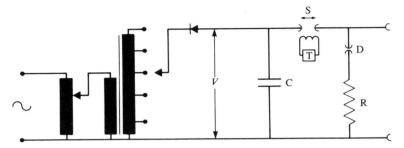

Fig. 4.1 The basic construction of a modern surge generator

capacitance C is usually selectable between a few microfarads and over a hundred microfarads by series, parallel and series-parallel switching, this selection being linked with voltage tappings and suitably interlocked. The arrangement is designed to give a constant energy output either with low voltage/high capacitance or high voltage/low capacitance.

The energy available from such a circuit is given by

$$\frac{1}{2}CV^2 \quad \text{joules or watt-seconds}$$

where C is in microfarads and V is in kilovolts.

When the stored energy in the capacitor is discharged into the cable, a very high energy steep fronted surge wave is launched, seeing an impedance of Z_O ahead of it. Examples of such voltage impulses are shown in Fig. 4.2. This is a series of curves with axes; V (kV) and t (μs). The heavy line represents a typical fault characteristic which shows that a fault will never break down instantaneously and that it takes a certain minimum voltage to break it down.

It is a common misconception that a fault must break down if a surge of a high enough voltage is applied to it. This is not so. A surge is not a steady d.c. potential but a wave with a very steep front edge and an exponentially decaying trailing edge as shown.

It should be clear from Fig. 4.2 that, while surge a has a high enough peak voltage, say 15 kV, for its tail to cut the fault characteristic at time t_1 and thus cause breakdown, it is possible to send a wave like b with a peak voltage of, say, 10 kV, that does not cut the fault curve and therefore does not cause breakdown. A voltage of 10 kV may well be above the already established withstand voltage of the fault but the problem is that not enough capacitance is in the circuit to ensure a slow enough decay of the wave. The curve c is that for a discharge at the same voltage as b but with more capacitance selected. This curve cuts the fault curve at time t_2 and causes breakdown.

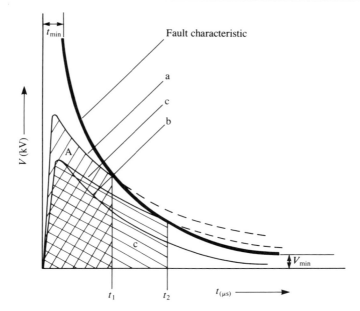

Fig. 4.2 Comparison of surges with different voltages and energy contents

An important point now becomes clear from these examples: more energy is dissipated by wave c than by wave a. In physics power is something waiting to be used. Energy is power being used; hence time is a major factor as evidenced by the units of energy: kilowatt-hour, watt-second. The energy under curve c is greater than that under curve a, which serves to illustrate that it is often better to increase capacitance primarily to create a breakdown and then to enhance the energy dissipation at the fault and thereby to create more noise/vibration.

In air, this is remarked as a very loud report or 'bang'. The cable, however, is usually buried at a depth of half to one metre and therefore the sound is muffled, producing a 'thump', which can sometimes be heard unassisted overground but more often can only be detected using a sensitive ground microphone and receiver/amplifier. Because of the noises they produce surge generators are often called 'bangers' or 'thumpers'.

As stated, this process is repeated every few seconds so that the operator can walk over the suspect zone to pin-point and mark the location exactly. Different people have different ideas on the best rate of discharge to use. Most go for a rate of one surge every three or four seconds. If the rate is over about ten seconds, it is possible to walk over the site of the fault without hearing the discharge. On the other hand, a rate quicker than approximately one surge every second does not give the d.c. source enough time to charge up the capacitors to the selected voltage before they are discharged again, thus putting an unnecessary limit on the energy available, remembering that this is proportional to V^2.

The features normally fitted to commercially available sets are:

- Safe enclosure
- Protected terminals
- Automatic 'dump' on switch-off
- Interlock to prevent high voltage appearing on switch-on if voltage control is not at zero.
- Selectable voltage and capacitance, interlocked
- Switch rate selector
- Voltage indication

Features that may or may not be included are:

- Dump to earth between surges
- 'Single shot' facility
- Use as HV test set
- Possibility of using HV with capacitors connected
- Remote control.

Generally, two sizes of surge generator are available: a smaller, portable type of unit for use on low voltage, control, indication and (some) telecommunications cabies, and the larger set in more common use, which is transportable, weighing up to around 100 kg. The smaller versions tend to have a capacity of a few hundred joules and an upper voltage limit of approximately 5 kV, while the larger models can deliver surges of up to 5000 + joules at voltages up to about 30 kV.

Some instruments currently available are shown in Fig. 4.3a to c. There are also rigs that can surge at 70 or 80 kV but, of necessity, these are modular arrangements mounted in the protected rear zone of a test van.

Given that a surge generator has been connected to a faulty cable, it is not good practice simply to switch on and rush up the route to listen. Successful pin-pointing depends on:

- Vibration/noise *actually* being created at the fault site.
- The operator being *certain* that this is so and
- *Standing over the cable* while listening

These points may seem simplistic but they are absolutely vital. These will now be taken in turn.

ACTUAL NOISE OR NOT

The surge generator may be functioning perfectly at the test end but there

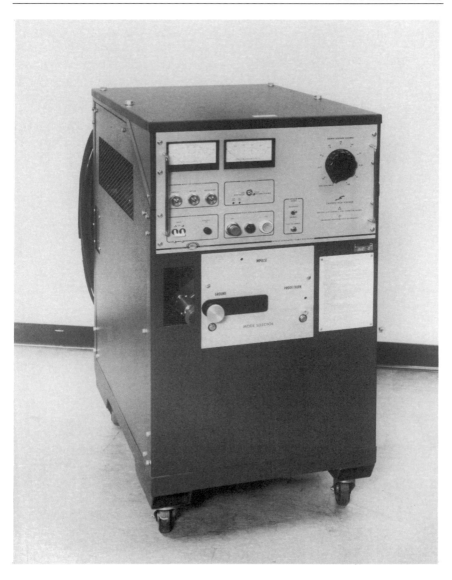

Fig. 4.3 (a)

Fig. 4.3 (a) The 30 kV, 5400 J heavy duty impulse generator with 65 kV test/burn (*source:* AVO Biddle Instruments, USA). (b) The T219 2.5/5 kV, 375 J surge generator with discharge between shots (*source:* Biccotest Ltd, UK). (c) The SWG 1000C 32 kV, 1000 J shock discharge generator (*source:* Seba Dynatronic GmbH, Germany)

118 UNDERGROUND CABLE FAULT LOCATION

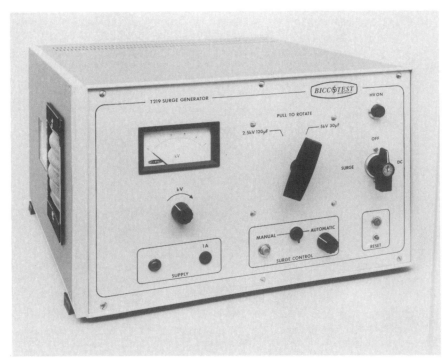

Fig. 4.3 (b)

Fig. 4.3 (c)

will be no noise at the fault if it is a zero ohm short circuit or if its resistance suddenly changes and the chosen voltage does not regularly break it down. It pays, therefore, first of all to measure the fault resistance immediately before starting the surge generator and then to surge for a few minutes, switch off, disconnect and measure the fault resistance again. If the resistance has remained at a medium or high value or has risen from zero or a few ohms to a medium value of, say, a few kilohms, it is certain that noise is being created by the surging. If impulse current equipment is also connected at the time, evidence of ionization delay proves that arcing is taking place.

In any case, it is sensible to wait for several minutes to make sure that the surge generator is being discharged regularly into the fault. As already stated, a fault can be unstable and give rise to intermittent discharging which is evident when the surge generator voltmeter does not kick back towards zero after each shot. If such a situation exists, the surge voltage can be raised so that the fault discharges every time. A good tip here is to switch off so that the core is earthed before surging is restarted at the higher voltage.

For instance, if 10 kV is not breaking the fault down and the voltage is increased by 2 kV, the next surge is 2 kV on top of the 10 kV already standing on the core. This may be doubled by reflection from the far end, thus stressing the fault at 14 kV. If, however, the core is discharged, a new surge of 12 kV will be doubled to 24 kV which is much more likely to break the fault down. Even if the fault will *still* not cooperate, at least the operator knows to spend longer at each point when listening over the cable so as not to be misled by the occasional surge producing no noise.

CERTAINTY

While the foregoing precautions will contribute greatly to the operator's certainty that the surge is producing a noise, there still remains the

120 UNDERGROUND CABLE FAULT LOCATION

possibility that the surge generator has ceased to operate. This may be due to a power failure, the surge generator having been inadvertently switched off or the fault having gone permanently 'high' so that the voltage is simply standing on the cable core and there are no breakdowns.

There are two remedies for these situations. The first is, perhaps obviously, *communication*. The surge generator may be several kilometres away from the suspect zone but someone should have been instructed to watch the surge generator carefully and to notify the field operator immediately should any of the above problems occur.

The second remedy can be provided by a facility built into some receivers/amplifiers. This is a pick-up coil and associated meter/indicator which indicates the presence of the magnetic field resulting from the surge current. This is an invaluable feature which, fortunately these days, has been incorporated by most manufacturers.

STANDING OVER THE CABLE

It is patently obvious that a fault can never be detected unless the microphone is placed directly over the cable. However, it often happens that the operator is listening for a discharge where there is *no cable*! It is a great shame if a perfect pre-location is made and then insufficient care is taken over route checking, tracing and marking. No matter how long the cable route it is normally easy to inject an audio frequency signal and confirm the cable location in the suspect area. The cable should be traced and marked every two to three metres over a distance of approximately 2 per cent of route length on each side of the calculated pre-location.

Figure 4.4 depicts a ground microphone with associated amplifier and headphones standing on the ground directly over a cable fault. The energy suddenly released at the fault site on the arrival of each surge wave produces a low frequency 'thump' as the shock waves travel through the ground cover to the surface.

The loudness of this and the ease with which it can be perceived vary tremendously. Occasionally it can be heard unaided by someone standing tens of metres away from the site. Sometimes it can be felt through the soles of the feet. More often it cannot be heard at all without the aid of a ground microphone and amplifier. Such a listening kit comprises:

1. A ground microphone, usually of the dynamic type, encased in a heavy cylindrical metal housing and having three legs and a handle. There are some crystal microphones but their use is decreasing. Some manufacturers offer a larger, heavier, wind protected model which stands on a stiff rubber or plastic skirt which protects the sensing element from wind that can otherwise mask the wanted sound or blot it out altogether.

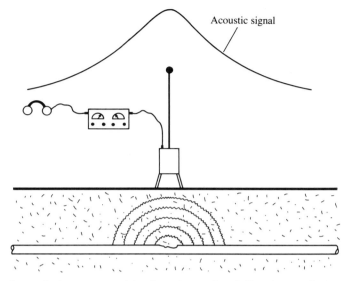

Fig. 4.4 Acoustic detector set with microphone placed directly over fault

There is a shielded lead to the amplifier and a 'mute' button mounted on the microphone handle or on the amplifier itself. This is necessary to damp down the ear damaging noise that will otherwise occur when the microphone is moved.

2 An amplifier which is carried on a strap around the neck. This may be a dedicated instrument or it may double as the receiver in an audio frequency set, depending on the manufacturer. The amplifier will incorporate an output to headphones as well as visual indication of the acoustic signal together with combined or separate volume/level controls. There may also be frequency band selection. As mentioned earlier, a feature of many such amplifiers is the detection and visual indication of the *magnetic* signal and, in some cases, there is a facility for measuring the time lag of the acoustic signal (see Fig. 4.5 and associated text).

It is a sad fact that, because the listening set must be very sensitive to the low frequency vibration being listened for, it is a superb instrument for picking up traffic vibration, the click of heels on a pavement, the movement of nearby objects and the thump and rumble of machinery. If the operator does not know the timing and rhythm of the surge generator, sorting out the wanted noise from the unwanted can be very difficult and often impossible. However, the human brain and senses are a formidable combination in detecting the coincidence of disparate signals, in this case visual and aural.

122 UNDERGROUND CABLE FAULT LOCATION

The well-known 'cocktail party effect' explains how the small sound produced by a fault can be identified in the much louder mish-mash of interference by the tuned ear of the experienced operator, particularly as the eye is indicating when the wanted sound should occur.

3 Headphones. These must simply be high quality, properly matched and very well protected against extraneous noises.

BEST PROCEDURE

Good practice in listening for vibration often marks the difference between success and failure in pin-pointing a fault. The first thing to be done is to check that the ground microphone, amplifier and headphones are connected up and working correctly. Before moving to the suspect zone, the set can then be positioned over the cable near to the test end for the operator to check that the magnetic signal indicator is working, set levels and generally get a feel for the timing of the surges.

The gear can then be moved to the suspect area where, as emphasized previously, the cable route will have been marked every two to three metres each side of the calculated location for a distance of between about ten and several hundred metres, depending on the total length of the cable. The operator should then place the ground microphone at the point calculated for the pre-location in the (normally vain) hope that the fault will be exactly there. Even though there will probably be no sign of the fault at this point, the operator should wait a little while, getting an idea of the surge rate, adjusting levels, filters, etc., and noting the type and extent of interference in the locality, e.g. wind, traffic, grass, pedestrians, etc.

At this point it may be reasonable to go to a point nearby where there are genuine grounds for suspicion, such as a recent excavation. However, if there is no such circumstantial evidence in the suspect area, the operator should then move off in one direction or the other, from the theoretical location, re-positioning the microphone and listening carefully at intervals of approximately half a metre. Many operators take several paces between listening points, but this is a mistake.

Although a high percentage of faults begin to be heard several metres away, it is not uncommon for the vibration to be confined to less than a one metre circle. Sound is best transmitted through well-compacted, homogeneous ground which has not been disturbed for a long time. If the cover over the cable is loose and faulted or made up of sand or stones, the noise is grossly attenuated and can be very difficult to hear.

The microphone (muted!) must be placed firmly on the surface so that it does not make loud noises by settling and the operator should listen very carefully for the duration of several surges just in case the fault is arcing over intermittently. The actual sound of the fault is similar to that produced by tapping a finger on the side of a wooden desk and the operator can soon acquire an ability to recognize this typical sound after hearing it repeatedly in different environments and circumstances.

Supposing a certain distance has been covered in one direction and no fault has been detected, a decision must be made as to how much farther to progress in that direction before starting to check in the other. As a general guide, about 1 per cent of the route length should be covered in one direction before returning to the pre-location mark and trying the other direction. This is not the case, of course, if the cable is very short, say only a few hundred metres, where poor route measurement or knowledge of loops and termination runs will often account for more errors than instrument inaccuracies. In these cases it is worth checking five to ten metres in one direction before going the other way. A clear mark should be made at the last position checked in case it is necessary to check further in that direction; this avoids a gap being left when restarting. The fault could easily lie within this missed stretch and the whole fabric of a logical thought process would be destroyed.

When the sound is eventually heard, its actual source at the fault site will probably be a few metres away. It is then a simple matter to listen further along the marked cable route and to each side of it to establish the exact point at which the sound is loudest. The visual indication of the acoustic signal may also be useful here. In most cases there will be no doubt at all of the location, and the ground can be excavated and the fault exposed.

124 UNDERGROUND CABLE FAULT LOCATION

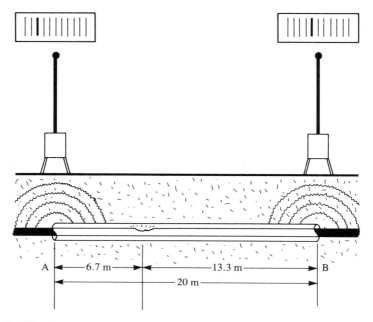

Fig. 4.5 Time lapse measurement for locating fault in ducted cable

There are some circumstances, however, in which it is impossible to be sure of the exact site of the fault even though it can be heard. The most common of these occur where the cable runs through ducts or under a solid raft of concrete. When a cable fault lies within a ducted stretch the sound of the discharge can be loudest at the ends of the duct, whereas a concrete surface tends to sound all over. In these cases, the exact fault location can only be found if the receiver/amplifier is equipped with time measuring circuity which provides a visual indication of the time difference between the arrivals of the magnetic and acoustic signals.

Where both signals are clearly present, such a receiver will indicate a decreasing time lag as the fault is approached, falling to a minimum directly over the fault. The lag increases again as the microphone passes beyond the fault. Where both signals are not necessarily present, as is often the case over the middle of a duct as the fault may not be audible, the time lag measurement facility can be used in a quantitative way to calculate the fault position within the pipe.

Figure 4.5 shows a cable in a duct 20 m long. The noise audible at end A produces a visual reading of 2 and that at end B a reading of 4. The fault is clearly nearer to end A and the actual distance from end A can be calculated approximately as

$$\frac{\text{End A reading}}{\text{Sum of readings}} \times \text{length of duct}$$

i.e. the distance to the fault from end A is

$$\frac{2}{2+4} \times 20 = 6.7 \text{ m}$$

4.2 Electromagnetic

This method involves checking at points along the cable for the *presence* and *direction* of the electromagnetic signal created by the current produced by the steep fronted voltage impulse when surging. The equipment comprises a pick-up coil feeding a receiver with a polarized indicator which gives the magnitude and direction of the signal. The pick-up coil is marked so that the same end is always pointed to the source to make sure that directional information is correct.

There are certain limitations and conditions associated with the use of the method, the most important of these being that it is confined to ducted systems, because the pick-up coil needs to be placed directly on the cable itself. Having said this, it can be used successfully providing that the operator fully understands the flow of currents in the particular cable system that is being worked on.

With unsheathed cable cores there is, of course, no difficulty in picking up a good signal. However, it is used mainly on high voltage cables that have metal sheaths or shields and, before moving on to the distribution of currents at various points along a cable, it is essential to discuss the *resultant* currents and magnetic fields in and around shielded cables.

Figure 4.6a shows a truly concentric situation with the core lying exactly in the centre of the annulus formed by the sheath. The *go* current in the core, shown as positive, generates a concentric magnetic field around the core and the *return* current in the shield, shown as negative, also produces a concentric magnetic field around the core, its notional centre of origin. Figure 4.6b shows the *go* current in one core of a normal multicore cable in which the cores twist within the shield and therefore adopt different, eccentric, positions at different points along its length.

Referring to Fig. 4.6a, it is clear that the *resultant* magnetic field surrounding the cable is produced by the *sum* of the *go* and *return* currents, this being a *zero* signal if the currents are equal. Likewise, if the *go* current predominates, the resultant will be in the positive direction and vice versa. Furthermore, a coil placed at N, S, E and W will pick up signals identical in magnitude and direction.

In Fig. 4.6b, however, while the *return* current in the sheath produces a concentric field, the point source of the *go* current in the core is *eccentric*.

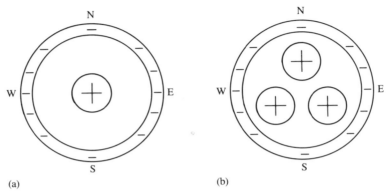

Fig. 4.6 Go and 'return' currents in (a) concentric cable and (b) three core cable

Therefore it is not possible to sum the two currents arithmetically as their relative positions have an obvious bearing on the resultant. For instance, if the coil is at N and the *go* and *return* currents are equal, the resultant field will be definitely positive. When the coil is placed at S, however, the field from the *go* current in the core will be less dense because its source is now farther away, and its interaction with the *return* current field results in a signal that is smaller and positive, zero or even negative, depending on the current magnitudes and the dimensions of the cable.

Figure 4.7 represents the normal situation found in practice, that of a multicore cable run in ducts, showing two manholes at C and D with a fault lying between them at F. The manhole at C represents those before the fault and contains an earth bond. Similarly, manhole D, also the site of an earth bond, represents those after the fault. The faulty core only is shown, it being assumed that the other cores are not involved in the fault and have not been earthed down at any point and thus carry no surge currents. The *go* current shown in the core flows to the sheath at the fault F and, as on most normal networks, returns to the surge generator via the

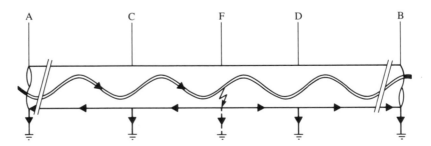

Fig. 4.7 Currents in a multicore cable run in ducts with manholes

sheath and other paths made up of earth bonds, spurious contacts between sheath and earth and, occasionally, from the faulty core and earth when there is a hole in the sheath at the fault. As already explained, the core shown will be eccentric within the metallic sheath at all points along its length.

A point not yet mentioned is that the electromagnetic method is equally effective whether or not the fault is arcing or is a short circuit, as current obviously flows in both cases.

To trace the fault the surge generator should be started up and checked for correct operation, i.e. it should be discharging consistently into the fault. Once the surge generator is running, the detection equipment should be checked out and levels set in a manhole close to the test end A, i.e. *before* the fault. (If no eccentricity is noted the fault may well lie between the test end A and this manhole.) It is important that these level settings are not changed till the test is completed.

The pick-up coil should then be applied to the outside of the cable at manholes along the route. In manholes lying *before* the fault, e.g. C, a definite *eccentric effect* should be evident whereby a maximum positive signal is detected at one point on the sheath associated with a minimum positive or even a negative signal at a diametrically opposite point, as there could even be a change of polarity, as already explained in the text referring to Fig. 4.6b. This variation of signal around the sheath means that the fault lies further along the cable as there is still *go* and *return* current in it.

If there is an earth bond in the manholes it is worth checking it for signal (with the 'source mark' on the coil pointing to the cable). This bond current should read positive and its value will be greater the nearer the manhole is to the fault. All readings should be noted down.

In the manholes after the fault, e.g. D, the readings around the circumference of the sheath will be uniform but still positive because, although there is no *go* current in the core, the concentric field is produced by the current in the sheath which is now flowing away from the source. There is a situation where the eccentric effect is detected both before *and* after the fault. This is when the fault path takes a significant time to ionize and the front edge of the surge wave passes beyond the fault, the breakdown occurring at some point on the 'tail' of the surge. Therefore, as shown in Fig.4.8, the wave has been chopped and the ongoing front portion of the wave will create an eccentric field after the fault. If this situation should arise, raising the surge voltage can reduce or eliminate the effect.

It is usually possible to identify manholes before and after the fault, enabling the faulty section between them to be replaced. However, how can the luckless operator decide this? Only perhaps after traversing the whole cable route and getting inconclusive results!

128 UNDERGROUND CABLE FAULT LOCATION

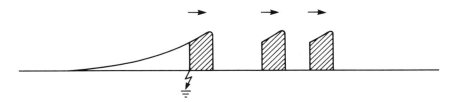

Fig. 4.8 Wave chopping caused by ionizing delay

Problems are also encountered when tracing faults on single core cable systems because there is no eccentric effect. A uniform positive indication will be noted both before and after the fault. There may be different amplitudes on each side of the fault but the change need not be significant.

A ploy that can be effective in this situation is to use a strap to form a low ohmic shunt between two points on the metallic sheath. The pick-up coil is then applied between these points and will show a strong positive indication at manholes before the fault, whereas a much weaker positive signal will be produced using the same technique at a manhole after the fault.

This section on electromagnetic fault tracing was prefaced with the caution that it had certain limitations. The foregoing serves to emphasize this. Furthermore, experience and knowledge gained from the impulse current method over the last decade indicate that long ionization delays are much more common than had previously been realized, thus increasing the incidence of wave chopping. This, together with the fact that prolonged thumping can involve more than a thousand surges, makes this method of fault tracing even less attractive on the grounds of cable stress and surge generator contactor wear. These uncertainties and disadvantages conspire to promote what is good practice anyway – the use of pre-location whenever possible.

Finally, it could be said that perhaps the best application of the electromagnetic tracing method is in pin-pointing very low resistance short circuits. In such cases it is easy to make a pre-location so that only a few manholes need be checked and the method comes further into its own as there will be no discharge sound to listen for.

4.3 Pool of potential

This method can be used when a cable is directly buried, i.e. in physical contact with the mass of earth at all points along its length and the faulty conductor is exposed to and touching the ground at one point only. The term 'conductor' is used advisedly. It may be the normally insulated metallic sheath of any type of cable. Equally, it can be the core of any type

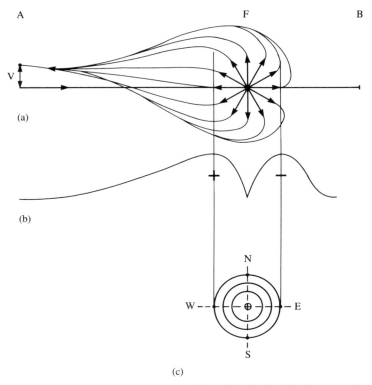

Fig. 4.9 Pool of potential – current and voltage distribution in the ground

of cable which has no metallic sheath. In other words, there has to be a length of metal insulated from the surrounding earth with damage to its insulation at one point, allowing galvanic contact with the earth as shown in Fig. 1.2a and b.

A voltage is then applied between this conductor and a ground spike or general earth. This produces currents in the ground whose paths fan out from the point of contact as shown in Fig. 4.9a to c, a and c thereof being plan views.

The steepest voltage gradients occur along the high density radial current paths within the 'pool', as shown in Fig. 4.9b. Figure 4.9c shows the 'pool' as a series of concentric equipotential circles which indicates, for instance, that the measurement W–O could be a high positive voltage and O–E would be a similar negative voltage. W–E would give a zero reading. This is also true for the N–O–S line and indeed for *any* diameter across the pool. In practice the gradient on the side away from the source tends to be less marked than the one on the source side so the lines of equipotential are not actually circular.

To locate the fault two ground spikes connected to a sensitive voltmeter are used to sample these voltages and find the point of zero potential at the centre of the pool. There are variations of this method to be discussed later that use alternating quantities, but the foregoing can be taken as representing the d.c. or instantaneous a.c. situation.

PRACTICAL APPROACH

The equipment needed for a field location consists of:

- A high voltage d.c. source
- A centre zero voltmeter with switched ranges
- Two sturdy ground spikes
- Two connecting leads, one long, one short

Figure 4.10 shows the MAGPIE set which includes a magnetometer for the magnetic field method (see Sec. 4.4).

High voltage d.c. sources are available with voltages from 2 to 12 kV. Some are dedicated units while others double as small burn-down sets, but all are 'short circuit proof'. The basic facilities provided are:

- Voltage control
- Output indication
- Continuous and interrupted output

The centre reading volmeter must have switched sensitivities so that it can be used to indicate millivolts or hundreds of volts. An essential facility is a zero offset control sourced from an internal battery. This is required because there are always currents, and therefore potentials, in the ground which cause serious deflections from the centre zero mark, particularly on the more sensitive ranges. The internal offset circuit can be used to trim the needle back to zero when it wanders off due to these stray ground potentials.

The ground spikes, about 1.25m long and of stout construction, should have insulated handles at the top and provision for connecting the wires from the voltmeter. Welded-on steps for forcing them into the ground are also useful.

Returning to the transmitter, i.e. high voltage d.c. source, the reason for having the interrupted output facility is to overcome the difficulties caused by the very ground potentials just referred to. If a steady voltage is applied to a ground contact fault and the voltmeter shows a deflection when the spikes are stuck in the ground at two points it is usually impossible to tell if the deflection is due to the applied voltage or a ground potential. To make matters even more difficult, the voltmeter needle will constantly shift as the stray potentials vary. It is therefore necessary to switch the applied voltage

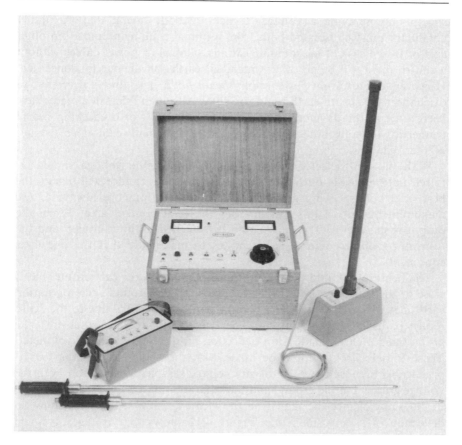

Fig. 4.10 MAGPIE set (*source:* Biccotest Ltd, UK)

on and off so that the operator can differentiate between the deflections produced by the transmitter and those due to stray voltages. Moreover, the output switching must have an *asymmetrical mark–space ratio* such as one second on, two seconds off. If the switching is regular, it is impossible for the operator to tell if a deflection means on or off.

A pool of potential test can be carried out by one person but is much more convenient with two. One carries a ground spike only while the other carries a spike and the volmeter on a strap around the neck. This person interprets the readings and directs the operation generally.

The test is commenced by connecting the transmitter to the faulty core or sheath and applying a voltage. It goes without saying that the faulty conductor must be isolated both from terminal equipment and from other earths. In the case of telecommunications, control or power cables without a sheath there will be no other (intended) earths on the core(s) concerned. However, when testing power cable *sheath faults* it is always necessary to disconnect the sheath earth bonds or links. Although this will already have been done at the diagnosis stage, it is nonetheless worth checking before connecting the transmitter. Earth contact faults normally exhibit a resistance of a few kilohms.

With the unit selected for continuous output, the voltage should be raised until a steady output current is flowing. Experience will dictate the best level but, speaking generally, it is not good practice always to use maximum output as this can dry out the fault contact area. When the output is satisfactory the unit should be switched to 'intermittent' and the transmitter observed for a few minutes to make sure that the output is consistent.

The operators should then move out to a point on the cable route that is close to the source without being very close to the ground spike or station earth mat where there will be a pool or pools of potential due to the dense return paths of the test current. The two spikes are then stuck into the ground and a check made for a deflection on the centre reading voltmeter. These initial checks should be made with the spikes as far apart as is practicable because the value of any voltage sampled depends on both the voltage gradient in the ground and the separation of the spikes.

Unless by chance the fault is in the vicinity, the first tests will show no deflection or a very small one. Figure 4.11 shows that voltages sampled before the fault will be of one sense and those after the fault of the opposite sense. The convention chosen is to call those before the fault positive and

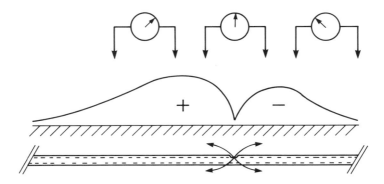

Fig. 4.11 Voltages sampled before and after the fault

those after the fault negative. While carrying out a test there is no particular need to be specific regarding polarity. The transmitter will normally have a negative output and the voltmeter may have positive and negative markings, but the apparent polarity of a deflection will ultimately depend on which way round the voltmeter is connected to the spikes and which spike is nearest to the source.

It is important only to adopt a convention for the test and to adhere to it throughout. For example, one operator should always be nearest the source and the first deflection noted should be recorded as 'left' or 'right'. As the operators move on up the route the spikes must then be kept in the same order. Thereafter, any change of sense on the voltmeter means that the fault position has been passed.

After the first check the operators move away from the source and stick the spikes in again, keeping the distance between them about the same. Again, there may be no deflection or there may be a slightly greater deflection in the same direction – say right. Once a deflection is noted the operators should move up the route, making checks at more or less regular intervals and keeping the spike separation more or less the same. The voltage will increase at each step and the voltmeter sensitivity will have to be decreased accordingly.

The distance between these checks will be anything from a few metres to a few hundred metres, depending entirely on the route length and the environment, e.g. open country, dense forest, roadway, verge, pathway, urban area, city centre, etc. If the cable is many kilometres long it will probably be necessary to use vehicles to 'hop' to various accessible parts of the route.

It should be remembered that no pre-location will have been possible on most ground contact faults. Therefore there is no alternative to surveying the whole route. There are occasions, of course, where a pre-location of, say, a core-to-core situation has been made on a cable with no sheath and the cores are in contact with the mass of earth at the same point. It is then an easy matter to home in on the fault.

In any case, the purpose of the exercise is to detect a change of polarity. Once a deflection to the (say) left has been detected, it is certain that the fault lies between this (left) test position and the last position back towards the source where a deflection to the right was noted.

Unless it has been necessary to 'hop' along the route, the operators will have noted ever-increasing deflections to the right, then decreasing right deflections followed by increasing left deflections. This pattern occurs as a maximum positive (right) voltage is indicated when the most advanced spike, i.e. the one furthest from the source, is just short of the fault position. As the probes are then moved onwards this most advanced one

134 UNDERGROUND CABLE FAULT LOCATION

begins to sample an ever-increasing negative (left) voltage. The voltmeter, which reads the difference between the two voltages, will therefore progressively show a decreasing positive reading and then zero when the probes are equidistant across the fault and, finally, ever-increasing negative readings as the rear probe advances towards the fault position.

As this process is carried out it is necessary to switch to less sensitive ranges and to decrease the span of the spikes. It is quite easy to see roughly where the 'zero position' lies; therefore it soon becomes possible for one operator to handle both spikes and to pin-point the exact zero point, which should then be marked.

The test so far has been carried out only in one direction, i.e. along the cable route. A check should now be made along a line perpendicular to this, such as CD in Fig. 4.12, which is again a plan view of the cable route. Note that the test route AB is shown slightly to one side of the cable route. It is normal to follow the cable with the AB test but it is not necessary to be directly over the cable. In fact, the cable route has no bearing at all on the current paths. The currents simply flow from the fault, the point of ground contact, back to the source ground spike or station earth by the best paths they can find. They will be evenly distributed in homogeneous soil but concentrated on any spurious low resistance path such as a metal pipe. From the first mark X on the AB path a perpendicular line should be marked. Sticking the probes in along this CD line will indicate the true 'zero voltage point', i.e. the point of ground contact.

Pin-point accuracy is possible with this method, the probe separation coming down to as little as 10 cm in many cases.

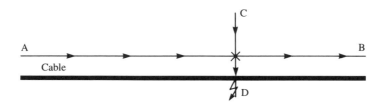

Fig. 4.12 Tests made along the cable route and along a line perpendicular to it in the locality of the fault

DIFFICULTIES – NO CHANGE OF POLARITY!

If the test is carried out as described above and no change of polarity is observed, it is probable that the fault lies at or near to one end or the other. If it lies at the far end B, a sharp increase in positive (right) indications will be noted close to this end and the cable can be confidently excavated at the B termination approach.

However, if the fault happens to be at end A or in the first stretch of cable close to A then it is possible that the test has been started further away from this section and very small deflections have been observed in one sense or, indeed, no detectable deflection at all. The fault therefore lies behind the starting point, i.e. nearer to end A, so it will be necessary to check back towards end A.

HARD SURFACES

Surfaces are often encountered, e.g. concrete, in which it is impossible to drive the spikes. In these cases there are several tricks that can be employed to overcome the problem.

Remembering that it is not necessary to follow the exact route of the cable, the spikes can be sunk into soft ground at points parallel to but off the actual route, such as cracks between slabs, adjacent gardens, verges, etc. This is also a solution when the cable runs behind a fence or other major obstacle. Alternatively, water can be poured on to the surface at the required contact points.

In the final analysis, holes can be drilled, but this is seldom necessary as it would then be worth while to employ an audio frequency method of pool of potential fault location using capacitive coupling (see Fig. 4.14 and associated text).

SEVERAL FAULTS

Although it is rare to have several faults on a cable core, it is far from unusual to encounter more than one point of damage on an insulated sheath. If regular routine maintenance is carried out, as is the case on medium and high voltage cables with a plastic oversheath to prevent corrosion of the (usually aluminium) metal sheath, either no fault or one fault is detected. This is then located and repaired. However, there are very many cables that are not checked on a routine basis which can have sheath damage at several points.

At the outset, of course, when the insulation resistance test is carried out and a low reading obtained, it is not known if this is due to one or more than one ground contact. When the d.c. transmitter is connected, the return currents may well emanate from several points along the cable. Thus several pools of potential will be formed but the operators will not know this as they set out to check the route. In fact, the several pools will probably exhibit different voltage gradients, the more severe ones occurring where large currents are leaking away.

However, the operators will detect and locate the first point of polarity reversal and expose and repair the damage. The straightforward re-testing of the sheath then shows continuing leakage and the route has to be

checked again. This process is repeated until the last damage has been repaired and the sheath withstands the mandatory pressure test (see, for example, Sec. 12.3 on methods of fault location, low voltage cables, sheath/armouring tests).

4.4 Magnetic field

The pool of potential method just described has been used very successfully for many years as a pin-pointing method. Faults are smug creatures, however, which do not give themselves up easily. It is also basically true that those which are easy to pre-locate are difficult to pin-point and vice versa.

This is certainly the case with the ground contact fault which has to be located by overground pin-pointing methods such as the pool of potential. There is a method, however, though not a pre-location method, which enables the fault location engineer to 'home in' on the fault quickly instead of having to traverse the whole route. The philosophy of this approach is basically that of the much maligned 'cut and test' method (see Chapter 9), in which a cable is cut at its mid point and each half tested. The faulty half is then cut half-way and the process repeated until the fault is exposed.

Instead of *cutting* the cable, in the magnetic field method a *directional* check is made at any point along the route that tells the operator that the fault is either *behind* or *in front of* that point. Therefore the search can be narrowed down to a small section relatively quickly. Then the pool of potential method is used to confirm the exact fault position.

This is accomplished by simply adding one more receiver to the pool of potential equipment, i.e. a magnetometer. Not much in fault location, however, is simple! The magnetometer is required to react to the varying magnetic field produced by the pulsed d.c. output current of the pool of potential transmitter. This current may only be milliamps but, for any chance of success, several hundred milliamps are required in order for the magnetic field produced to be strong enough for the magnetometer to detect. Even then this field is of the order of one thousand times less than the earth's magnetic field.

It is clear that not only must the magnetometer be specially designed but a modified version of the normal pool of potential transmitter is required. The modifications are twofold. Firstly, the power must be increased so that an output current of around 500 mA can be sustained into a typical ground contact fault. Secondly, as indicated later in 'practical approach', detection of the field is not easy, and so it is necessary to have some means of showing the cadence of the pulsed transmitter output at the receiver (magnetometer).

This is the same concept as the thinking behind the acoustic receiver used when surging, where there are two indicators, one showing the electromagnetic effect and the other the acoustic effect. The visible evidence of the occurrence of the surge tells the operator when the characteristic sound might be heard. Likewise, in the magnetic field method, some device is needed at the transmitter which 'tells' the receiver when the output voltage is on and off.

A galvanic connection is impossible over distance and a radio link possible but unnecessary as there is a simpler, though unorthodox, solution. This is to equip the receiver with a replica of the quartz crystal controlled timing device in the transmitter which is used to switch the pulsed output. This 'receiver timer' drives a visual indicator and runs at exactly the same rate and mark space ratio as the 'transmitter timer'.

Once the transmitter is set up and running, the two timers are simply synchronized. Thereafter, the quartz timing errors being so small, they will run together for hours if necessary. Clearly, this system is not absolutely watertight. The receiver is showing when the transmitter *should* be on and off. The transmitter operator could switch off and go home and the receiver operator would be none the wiser! However, in all fairness, this is a very neat solution and it is most effective in the field.

PRACTICAL APPROACH

The transmitter is connected up exactly as in the pool of potential method. In fact the pool of potential and magnetic field methods are one and the same except that the latter needs a higher output current and provides a means of locating the fault zone quickly.

As intimated above, the receiver is then switched on and the 'sychronize' button depressed and then released to coincide with the start of the transmitter timing sequence. This synchronizes the two units.

The receiver/magnetometer should then be observed for a short time to confirm that this is indeed the case, i.e. that the receiver indicator is moving in time with the transmitter output pulses. The receiver/magnetometer, the centre reading voltmeter and the spikes are then assembled and taken out to a point on the route close to the source substation or exchange/central

138 UNDERGROUND CABLE FAULT LOCATION

office and the receiver/magnetometer set down *directly over the cable*. Note that, for this test, it is necessary to place it more or less over the actual cable because it is sensing the standing magnetic field produced by the test current in the cable itself, not in the mass of earth. It should also be located away from vehicles and other large metallic objects such as massive fences or girders and observed for a few minutes to make sure that the pulsed indication is present.

It should be remembered that a magnetometer is an extremely sensitive piece of apparatus which is compensated for the earth's magnetic field. Despite this, rapid movements of the unit itself will produce flux linkages with the earth's field, so it should be placed on the ground and left still. Note that the presence of large masses of metal and particularly moving metal masses modifies the earth's field and produces deflections.

Having observed that the deflections are taking place at the correct rhythm, i.e. in time with the timer indication, the amplitude of the deflection should be set at about $\frac{3}{4}$ f.s.d. and the *direction* noted as to the right or to the left ... *with the arrow on the receiver/magnetometer pointing away from the source*. As with all directional tests involving inductive pick-up, the orientation of the pick-up coil (magnetometer in this case) is obviously relevant. After this initial check, *all* subsequent checks must be made with the receiver/magnetometer pointing the same way.

It is now time to start sectionalizing the route. Theoretically, the first check should be made half-way along the cable. Therefore the plans should be consulted and a point chosen which is near the mid point. In choosing the exact position for this, however, there are other criteria to consider which are more important than exactly bisecting the cable. These are:

- Ease of access
- Absence of vehicles
- Suspect areas
- Presence of joints, etc.

For instance, there may be a recent excavation nearby which might have caused the cable damage. It would be sensible in this case to select the first test site before this point. If the magnetometer indicates that the fault is still ahead of this position, the whole stretch behind the magnetometer is 'cleared' of suspicion and a short 'hop' past the excavation may well show the fault to be between the two test points, i.e. probably at the suspect excavation, a fact that can be quickly verified using the spikes and centre reading voltmeter.

Leaving aside such specific considerations, the logical routine to adopt is to check half-way along the cable as just stated, then half-way along the suspect half, etc., as indicated in Fig. 4.13. This shows the first check at

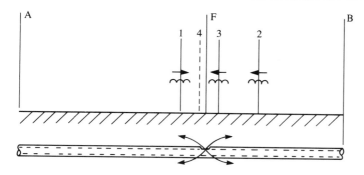

Fig. 4.13 Sectionalizing with the magnetometer

position 1, giving a deflection which indicates that the fault is still ahead. That is to say, the magnetometer deflection is in the same direction as the deflection obtained at the initial check near to A. There has been no change of direction.

The second half of the cable is now bisected and a test made at position 2. This *does* show a change of direction. Therefore the fault has been passed and lies between positions 1 and 2. This quarter length of cable is now bisected and a test made at position 3. Again the deflection indicates backwards, the same as at position 2, so the fault must lie between positions 1 and 3, a distance of one-eighth the cable length.

At this juncture there is a simple choice to make: either to bisect the remaining cable length at the possible position 4, shown dotted in Fig. 4.13, or to abandon the magnetometer approach and start a search as per the pool of potential method. This choice depends on the length of the cable.

If the cable is only 1 km long, the distance between positions 1 and 3 will be 125 m and pin-pointing by the pool of potential method can be carried out in a short space of time. Should the cable be, say, 5 km long, the distance between the same two positions would be 625 m and it is obviously advisable to further bisect this to produce a suspect zone 312 m long, or even again down to 156 m.

The 1 km long cable will therefore have required only three spot checks and the 5 km long cable, four or five. This approach is quite clearly preferable to walking the whole route and checking every 50 to 100 m with the ground spikes and centre reading voltmeter.

4.5 Audio frequency

Audio frequency methods of pin-pointing can be further subdivided into three sections:

- Pool of potential
- Twist method, core to core
- Twist method and others, core to sheath

POOL OF POTENTIAL

Pin-pointing by the pool of potential method has just been described whereby a d.c. voltage is applied to the faulty conductor which produces return currents in the mass of earth. In exactly the same way, an alternating voltage can be applied which duly produces alternating currents in the mass of earth. There is no basic difference in the test circuit, simply in the means of sampling the ground voltages. Figure 4.14 shows the ground voltage signal pattern.

The signal is produced by an audio frequency generator connected between the grounded sheath or core and a ground spike as before but, instead of a voltmeter, an audio frequency receiver/amplifier is used to detect the signal which is displayed on a meter or heard on headphones. Instead of separate ground spikes, a purpose built 'A' frame is used which is stuck into the ground by a sole operator. Some manufacturers also provide capacitor plates which can be fitted on to the legs of the frame for situations where the legs cannot be stuck into hard surfaces.

Just as in the case of the d.c. pool of potential method, a voltage is being sampled between two spikes (legs) but, in this case, the voltage is alternating and is reproduced as a tone in the headphones and as a signal amplitude on the receiver meter. The frequencies used vary between about 1 and about 12 kHz.

In the general consideration of this method we are dealing with signal magnitude and not with direction, although there does exist one variation that incorporates direction indication. This is covered in 'practical approach'.

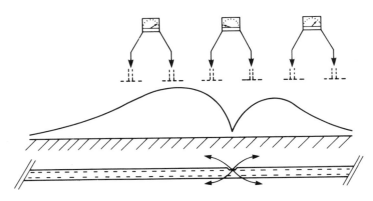

Fig. 4.14 Ground voltage signal pattern

The location is carried out by walking the route and listening at points along it. As the fault is approached there is a steadily rising signal volume which peaks just before it. As the frame legs straddle the fault the signal dies to a minimum but then rises quickly to another peak just beyond the fault, whereafter it gradually sinks to a low level as shown in Fig. 4.14.

Depending on the equipment being used, frequencies around 10 kHz are normally chosen, this being a necessity if the capacitor plates are being used. Signal levels are very high and indications quite distinct and sharp within a few metres of the fault, but are very low at points distant from it. As in the d.c. method, the current flow is in the mass of earth; therefore the equipment does not need to be directly over the cable.

PRACTICAL APPROACH

At the source cable termination, the faulty sheath or core is checked to make sure that no solid earth has been left connected. The audio frequency generator is then connected between the faulty conductor and a ground spike or general earth. The matching and amplitude controls should be adjusted for a maximum or near maximum output. Figure 4.15a to c shows some of the sets of equipment currently available. Although the photographs show audio frequency equipment for the pool of potential method, some of these sets include accessories for or are themselves part of sets designed also for cable and pipe tracing and, in some cases, cable identification.

It must be pointed out here that a detailed coverage of audio frequency induction methods in theory and practice is to be found in Part Two, Chapters 6 (route tracing) and 7 (cable identification).

When the audio frequency generator is set up, the current paths in the ground will be identical to those already described in the d.c. method and shown in Fig. 4.9a. The receiver/amplifier, together with headphones and 'A' frame are now taken to a point on the cable route close to the source and connected up. The transmitter frequency is selected and tuned in and levels of audible and visible signals set. It is probable that only a low signal will be heard at this point unless, by chance, the fault is in the vicinity.

The 'A' frame should then be advanced along the route and stuck into the ground every few metres and the signal levels checked. If the capacitor plates are being used, the signal levels will be even lower, but this is not too important as the survey can be carried out at a brisk walking pace over any terrain, and signal levels are good in the vicinity of the fault.

At some point along the route a small but noticeable rise in signal level will become apparent. This will develop rapidly into a very strong signal which has to be attenuated. Very suddenly, within the length of one pace, this will drop to a marked zero or minimum and, just as suddenly, rise to another crescendo before slowly dying away again.

142 UNDERGROUND CABLE FAULT LOCATION

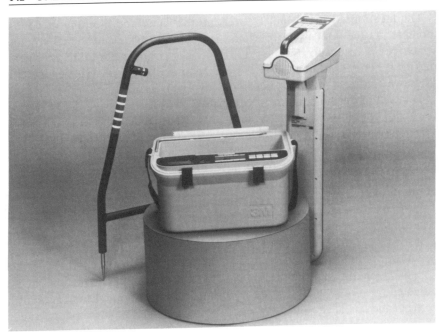

Fig. 4.15 (a)

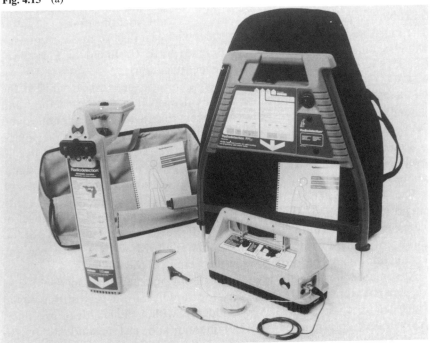

Fig. 4.15 (b)

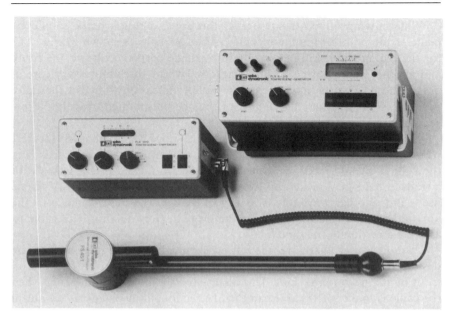

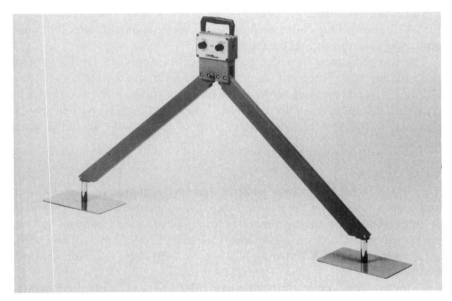

Fig. 4.15 (c)

Fig. 4.15 (a) The 'Dynatel' 573D cable and fault locator (*source:* 3M plc, UK). (b) The RD400/FFTx/FFL2 (*source:* Radiodetection Ltd, UK). (c) The 'Ferrolux' audio frequency set plus 'DEB' frame (*source:* Seba Dynatronic GmbH, Germany)

While moving quite quickly across this zone, a rough mark should be made at the point of minimum signal. Then, exactly as shown in Fig. 4.12 for the d.c. method, a traverse should be made perpendicular to the route and another minimum noted and marked. Two or three more very careful passes should then be made using the visual indication to detect the exact position of the minima. The fault can thus be pin-pointed to within a few centimetres.

As mentioned before, some manufacturers supply equipment for this test which has a directional capability. This is accomplished by synchronizing the voltage signal picked up from the ground via the 'A' frame spikes with a transmitted signal picked up by a search coil in the receiver or 'A' frame itself. Put another way, this is a way of giving a source reference to the alternating quantity sampled from the ground, thus making it possible to allocate direction to the signal.

On short routes, therefore, the test procedure will be the same as that just described, but with the added security of knowing when the frame is in front of and beyond the fault. This facility is even more advantageous when the cable is longer. The cable can then be sectionalized exactly as set out in the foregoing section on the magnetic field method and as depicted in Fig. 4.13. As with that method, there are safeguards to make sure that mistakes are not made by having the 'A' frame wrongly orientated. Either the 'A' frame legs are coded or the receiver meter is mounted on the frame and always points towards the fault.

The foregoing describes the audio frequency approach to the pool of potential method. There still remain other audio frequency methods of pin-pointing which are set out below, some of which will be referred to again or recapitulated in part in Part Three, 'specialized areas', particularly in Chapter 9. The first and most important of these is the so-called 'twist method'.

TWIST METHOD

This is the most well-known audio frequency technique in fault location and simply depends on the fact that the cores of a cable are laid up twisted. It dates back to the mid-forties when an all-valve audio frequency set was developed by BICC (British Insulated Calendars Cables Limited) and MEC (Midlands Electricity Company). It was named 'BIMEC' – a name that still exists.

From that time, on uncluttered paper cables, very accurate pre-locations could normally be made by experts using resistance bridges, but pin-pointing was a problem. Shock wave discharge had yet to come into general use but HV test sets and some (often improvised) burn-down sets were available and a common practice was to burn the fault down until a very low

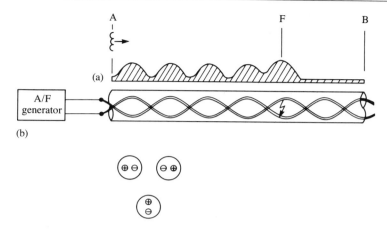

Fig. 4.16 Twist method

resistance path was produced between two cores. The twist method was then applied.

As shown in Fig. 4.16, the audio frequency (A/F) generator is connected between the two cores which are burnt together at the fault. For any reasonable chance of success, the core-to-core fault resistance must be near 0 ohms or at least under 10 ohms.

The audio frequency current passes down one core and returns via the fault path and the other core to the generator. At any point before the fault, therefore, there is a *go* and *return* current present in the cable, depicted in Fig. 4.16 as + and −. These currents produce magnetic field maxima and minima as shown in Fig. 4.16b at each half-twist of the cores. Therefore, if a search coil and receiver are moved along the cable, a series of signal maxima and minima is detected as shown. This pattern continues up to the fault position where a greater maximum is often detected. Beyond this point the signal tails off to a steady but very low level.

PRACTICAL APPROACH

The section of cable route to be walked over when pin-pointing a fault by this method will normally be short. This is because the cable concerned is a low voltage one whose length will rarely be more than a few hundred metres, or the search is to be made in the suspect zone of a longer MV or HV cable when a pre-location has been made. Again this involves a length of one or two hundred metres at most.

Before commencing it is essential that the route or route section has been traced beforehand and marked. The reason for this is that the twist method test will be carried out at a brisk walking pace to be most effective, and trying to trace the cable at the same time disturbs the recognition of the rhythm of the rising and falling tone pattern. The tracing can be carried out very quickly using the same set of audio frequency equipment as will be used for the twist method test.

Firstly, the resistance between the two cores concerned should be checked as still being under 10 ohms and, if the cable is a low voltage one, the neutral earthing links should be removed to reduce other possible signal paths. The audio frequency generator output is then connected to the two cores. The more power that can be applied, the better. Therefore, if a high power set is available, such as might be built into a test van, this should be used. It is likely that the normal portable battery/mains generator will be to hand; therefore it should be fed from the mains if at all possible in order to obtain maximum power.

The higher frequencies in the region of 30 to 80 kHz should not be used because of the capacity signal drawn. Best are the *ca.* 1 or 10 kHz frequencies. The *ca.* 1 kHz frequency is best if power output is good and the signal easy to pick up. However, a *ca.* 10 kHz signal is almost always clean and easy to detect. As always when using audio frequency for any purpose, the empirical approach is preferred. The operator should use whichever frequency gives the best results.

The matching switch in the generator should be used to select the position that gives the highest output and then the overall signal amplitude control turned to maximum. The operator then takes up a position at the very beginning of the route (or suspect zone), checks that the signal rises and falls and sets audible and visual signal levels accordingly. (If there is no rise and fall the fault is behind this position, very close to the source termination.)

Because the route is clearly marked, it can be followed at a fast walking pace. It is totally counterproductive to check slowly, wave the search coil about or keep stopping and starting. The feature being monitored is the rise and fall pattern of the signal; therefore it is best to walk fast and establish a steady rhythm of successive maxima. Any change or cessation will then be immediately obvious. The separation between signal peaks for power cables varies from around one third of a metre to one metre, so this fits in well with the human pace and walking speed. If the signal ceases or becomes steady this normally means that the fault is near the last peak discerned. However, it is as well to make further checks because this can also happen for other reasons:

1 The cable may 'dive' suddenly and rise to normal depths further on.
2 There may be an unrecorded tee branch up which the fault lies and the operator is continuing to follow the main cable route.

3 There is a straight joint.

In all cases the operator should walk on through this zone without hesitating while making a mental note of roughly where the last peak was.

The case of the straight joint is easy. After only a short dip in signal (or longer peak, depending on the search coil orientation), the rise and fall pattern immediately starts again. Indeed, this is one of the main ways of locating the position of a joint. This is covered in more detail in Chapters 6 and 7.

If the disappearance of the signal is due to a sudden increase in cable depth, the rise and fall pattern will be picked up again further along the route and there is no need to agonize over the signal loss. The fault *must* be further on.

The existence of an unrecorded tee branch, however, can be most awkward as no one will be expecting this to be the case. In this circumstance walking further along the route will not produce the rise and fall pattern again. Therefore, before making the mistake of excavating at the last peak position, the area should be circled and searched for the signal. This will very quickly indicate a rise and fall pattern along the previously unknown tee branch.

The above precautions having been taken, it is safe to mark the fault at the last peak.

The twist method is often the only way of pin-pointing faults on multibranched LV cables with loads connected, particularly, if the *ca.* 10 KHz signal is used. This is because the 'wanted signal' passes along the twisted cores to the fault and little signal is bled away by house loads and energy meter coils due to their impedance at this frequency. The method is therefore a favourite with hard pressed engineers engaged in urgent LV system trouble shooting. Further points regarding this application are made in Part Three, Chapter 9, 'power cables'.

The twist method technique cannot be applied to underground telecommunications and control cables because the ratio of their 'twist repeat interval' to their depth is too great. The maxima only occur every few centimetres and the rise and fall pattern is only discernible with the search coil directly on the cable or a little above it. However, the method is a very powerful one when the cable is accessible, as it often is in cable trays, troughing, when cleated to walls or at intermediate access points such as manholes and handholes.

For telecommunications and control cable fault locations (and cable identification), several manufacturers supply a small search coil about 3 or 4 cm long which can be placed on the cable and, if space permits, even moved around its circumference to check the two maxima and minima that occur around a cable carrying 'go' and 'return' signals in two cores. A low resistance short between two legs of a pair can be located exactly. It is also possible to check if the pair is peripheral or central by noting the disposition and intensity of the maxima around the cable.

Given these obvious differences between twist method applications on telecommunications cables and power cables, the approach is similar in all respects except that a lower power output can and should be used on telecommunications cables as the signal intensity is much greater with the pick-up coil situated directly on the cable.

4.6 Core-to-sheath-faults

In certain very specific circumstances it is possible to pin-point core-to-sheath faults using audio frequency methods. However, it must be stated quite emphatically that this is a very imprecise art. Generally speaking, any medium resistance fault (hundreds of ohms or kilohms) cannot normally be found. If the fault resistance is low, there is a chance of finding it with one of the three methods which are now described. The golden rule with all audio frequency fault location is, however, if no success has been achieved after about one hour ... abandon the attempt.

TWIST METHOD

This variation of the twist method gives the best chance of locating a core-to-sheath-fault. Figure 4.17 shows the test configuration.

The audio frequency generator is connected between a healthy core and sheath, but with this core shorted to the faulty core at the remote end. This means that the test current passes along the healthy core and returns to the sheath along the faulty core, resulting in a rise and fall signal pattern between the fault and the remote end. The signal can be checked as already described but this time starting from the far end.

It is recommended that, if at all possible, all sheath earths (and neutral earths on low voltage cables) are removed to eliminate other signal paths

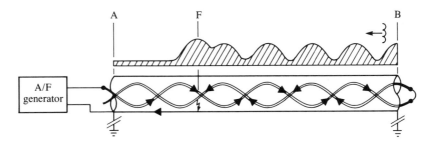

Fig. 4.17 Twist method for locating a core-to-sheath fault

which create magnetic fields that add to or subtract from the 'wanted' rising and falling field.

DOUBLE COIL ASSEMBLIES

Directly following on from this last point about interference fields, it is worth considering the use of the double coil assemblies that some manufacturers supply. This is particularly effective when using the twist method.

The coils are supplied mounted on a boom and their separation is adjustable. They are connected in 'opposition', i.e. if they sit in the same magnetic field, the signals induced in them cancel out. Conversely, if they are cut by different fields, there is a resultant signal. This connection has the effect of cancelling out interference fields such as those from adjacent metal lines and cables or, more importantly, from sheath and neutral currents in the cable being followed. Figure 4.18 shows a double coil assembly in use in a twist method situation.

The separation of the two coils is adjusted so that they lie over adjacent signal peaks. This is easy to do as the peak-to-peak distance can be ascertained with a single coil or, at the worst, from knowledge of the cable type and size.

As indicated in Fig. 4.18, the interference field produces the same signal in each coil as it is more or less constant around both of them. Suppose this signal is $+v$ in each. The signals derived from the twisted cores, however, will be 180° opposed as they are shifted by a *half-twist*. Call these signals $+V$ and $-V$. Naturally these are all complex quantities but, nonetheless, the principle of interference suppression can be illustrated simplistically by saying that the resultant signal perceived by the operator is

$$(V + v) - (-V + v) = 2V$$

Even without being scientifically specific it is clear that the 'wanted' signal is augmented and the 'unwanted' one reduced.

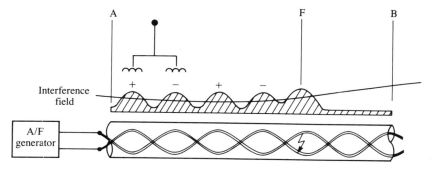

Fig. 4.18 Double coil assembly in use in a twist method situation

150 UNDERGROUND CABLE FAULT LOCATION

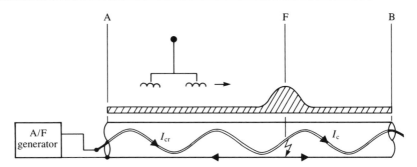

Fig. 4.19 Double coil assembly in use on a core-to-sheath fault

DOUBLE COIL ASSEMBLY – CORE-TO-SHEATH FAULTS
The double coil assembly can also be used on a core-to-sheath fault, as shown in Fig. 4.19.

When the two coils are situated before the fault, each produces a signal derived from I_{cr} (the capacitive drain of the whole cable plus the in-phase current through the fault). They are identical vector quantities; therefore their resultant is a theoretical zero – in practice a low, more or less steady, signal.

If positioned after the fault, they again produce a low resultant signal because each is generated by the field from I_C, the capacitive current drawn by the remaining cable length. However, when one coil is before the fault and the other one beyond it, they each produce a different signal, one derived from I_{cr} and the other from I_C, so that a peak signal is detected at this position.

This approach is only worth pursuing if there has been a good pre-location of a low resistance fault and the suspect zone is certain. If a good peak is detected within that zone, well and good. If not, or if there are several peaks, the test should be given up after all reasonable combinations of signal level, coil separation and orientation have been tried.

There are also pitfalls awaiting the fault location engineer using this method because several factors and features also produce peaks, for instance sudden deviations in the cable run and joints. Furthermore, the distances to the cable terminations before and beyond the fault obviously influence the levels of capacitive current at any point. To repeat the comment already made: if it works – fine; if not – try some other approach.

FIELD TURBIDITY METHOD
As shown in Fig. 6.3b in Part Two, Chapter 6, 'route tracing', a search coil placed in-line with a cable carrying an audio frequency current gives a minimum signal. If the coil is turned so that it is at 90° to the cable, a

maximum is detected. This is most useful when tracing cables in establishing the exact lay of the cable at any point.

These maxima and minima are noted whether the audio frequency source is being applied to an open ended core, a core shorted at the far end or a core with a fault on it. The magnetic field created is of the same form, although the current producing it may be in phase or at a leading angle.

However, in the case of a core with a fault to sheath, the magnetic field is often 'turbid' at the point of fault. That is to say, it is irregular and deformed, this being due to the fault path itself which is more or less perpendicular to the core direction. This generates an interactive field along a different axis to the axis producing the regular field recognizable up to and beyond the point of fault, where it often displays a much less sharp minimum signal whose 'bottom value' can be checked as shown in Fig. 4.20a.

The resultant magnetic field directly over the fault exhibits a different 'minimum signal axis'. The operator finds that, over the fault, a minimum signal is observed with the search coil not exactly in line with the cable run at that point. This is illustrated in Fig. 4.20b, which is a plan view showing the orientation of a search coil for a minimum signal as it moves along the cable over the fault.

It must be emphasized again that this, along with other audio frequency methods for core-to-sheath fault location, is useless on anything but low resistance faults. It can, however, be most effective on low resistance faults on short cables or over short suspect zones where shock wave discharge is ineffective or cannot be used. The lower frequencies are best here, particularly those under 500 Hz, if available.

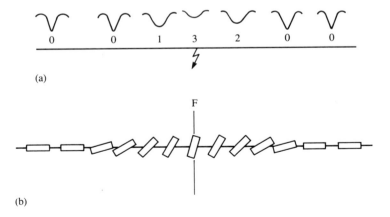

Fig. 4.20 Field turbidity method (a) variation of minimum signal (b) plan view showing the search coil orientation for minimum signal

5

Confirmation

5.1 Excavation, backfilling and reinstatement

At some point during the fault location it will become necessary to dig a hole to expose the cable. It may be that the person in charge is sure that the fault is located at a particular spot. It may also be that the hole is being dug on a suspicion that this is the case, or, indeed, just as a means of exposing a joint or for the purpose of making a test cut.

Whatever the reason for the excavation, there is a right and a wrong way to go about digging holes! The wrong way is simply to instruct the digging team to 'dig here'. The right way involves following certain simple but sensible rules which will avoid inconvenience and save time later. These are now given.

DECIDE THE SIZE OF THE HOLE
If the cable is a telecommunications or street lighting cable laid at a depth of about 300 mm, a hole 600 mm square may well suffice. However, to start a hole of that size when exposing a large high voltage cable laid 1 m deep is ludicrous – it will just about be possible to dig down to the cable, after which there will be no way of manipulating spades effectively. The excavation has to be restarted from wider boundaries.

Obviously the depth, size and type of cable influence the choice of hole size but the most important factor is *what has to be done to the cable once it is exposed*. In most cases it will be necessary to *cut* the cable. Afterwards, it will have to be rejointed and the new jointing may comprise one straight joint, one tee joint or several joints (if a new piece of cable has to be let in). Therefore, the content and span of the probable repair jointing are major factors in deciding hole size.

SAFETY NOTE

The sides of all deep holes should be shored up; otherwise workers are in great danger of being buried by a collapsing side wall. As a rough guide, shoring should be undertaken at depths over about 1.5 m but, in all cases, the regulations of the country concerned should be strictly adhered to.

TAKE CARE OF REPLACEABLE SURFACES

If the area is surfaced in paving slabs, cobbles or bricks, these should be lifted carefully without being damaged and stacked in an orderly fashion so that they can be replaced if at all possible in the positions they were taken from. Turf should be cut into squares and laid aside carefully with a view to its being kept alive and replaced in an orderly fashion when the hole is filled in. Note that if the fault is on a low voltage power cable it may have been burning for some time and there is often a characteristic smell in the soil around it which is most evident just as the slabs are lifted. The soil temperature will also be higher.

PILE THE SPOIL IN THE RIGHT PLACE

- *Not* on a garden causing damage
- *Not* where it will cause damage or inconvenience to foot or vehicular traffic
- *Not* on the stacked paving slabs or turfs
- *Not* over a side of the hole where further excavation may be necessary!

In practice it is amazing how often the spoil lies over the edge of the hole that has to be extended (see Fig. 5.1)!

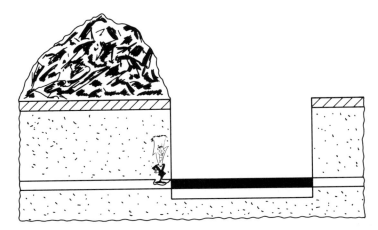

Fig. 5.1 Piling the spoil in the right place

CAREFUL DIGGING

It goes without saying that the utmost care should be taken not to damage the cable being exposed or any other cable or service encountered.

BACKFILLING

After repairs have been made it is very important to backfill the hole correctly. The following procedures should be carried out whenever possible:

1. Any damaged ducts, culverts or footings should be repaired.
2. Cables, joints and ducts should be supported underneath.
3. The first infill around the cable should be of sand or fine soil devoid of stones.
4. Cable protection tiles and/or marker tapes should be replaced.
5. The spoil should be tamped down to reduce subsidence later.

REINSTATEMENT

1. Slabs should be replaced in the positions from which they were taken up.
2. Turfs should be bedded properly and watered.
3. If the surface broken into is concrete, a temporary tarmac surface should be laid down.
4. No slab edge, hole, depression or irregularity deeper than about 2 cm should be left. This is meant as a guideline and not as a universal rule. The country, county, municipality or local authority concerned will state such limits in their laws and by-laws.
5. It is normal thereafter to notify the municipality in charge of the area concerned. For a charge they will then reinstate the disturbed surfacing.

5.2 Exposure

Having dug down to the cable, exposing the fault is quite routine, providing it is actually within the bounds of the hole excavated and providing there is obvious external damage.

If there are no signs of a fault as the cable is exposed, the cable should be cleaned off right around its circumference and all along its length. Particular attention should be paid to places where a sharp stone may have caused damage, but the whole surface of the cable should be inspected for any sign such as a dent, crack or hole, no matter how small. If this close inspection does not reveal the fault, then it must be that the fault is not in that section or is inside the cable, i.e. it has not been caused by external damage, nor has it blown the sheath open.

In the case of a telecommunications cable, any joint on the cable within the hole should be opened up as the fault will probably be in it. There may be no external sign, but water may have run down into the joint from damage further along the cable. Even if the fault is not in the joint, further testing from the joint will lead to a final location and the joint can be easily remade. Should the cable be a power cable, however, it would be best at this juncture to repeat the shock wave discharge test and use the ground microphone directly on the cable.

5.3 Analysis of cause

At all stages during the process of digging down to the cable, inspecting it and exposing the fault, thought must be given to establishing the *cause* of the fault. The causes that demand most care and attention are third party damage, poor cable manufacture and bad jointing practice. This is because some company or person is to blame and it should be possible to recover part or all of the fault location costs if liability can be proven.

Polaroid photographs should therefore be taken of anything that can help to establish liability such as a broken or collapsed duct line, missing protection tiles, a badly made joint, etc. The very nature of the investigation means that evidence must be destroyed in excavating, exposing and repairing the fault, so the taking of photographs is essential if any follow-up is to be made.

Other causes are worth establishing for technical rather than political reasons. For instance, the discovery of a faulty joint connecting paper and plastic insulated telecommunications cable sections may add to statistics which are used to decide what type of joints are best for that application.

5.4 Recording

All the points mentioned so far should be recorded, i.e. type of fault, accuracy of location and cause of fault, but, in addition to this, a sketch can be made. This should show the exact lie, disposition and depth of not only the cable under investigation but also other cables, pipes and services exposed in the excavation. The detail will then be added to existing cable records together with a final sketch plus notes on what cable and joints were used in the repair.

5.5 Further testing

Once a fault has been identified and a core or cores cut, it must never be assumed that the cable is then healthy. Further tests need to be carried out between this cut and the ends of the cable to check for faulty conditions.

If these tests show the cable to be 'clear', repair jointing can then be carried out, final tests made and the cable put back into service. Occasionally, though, tests at the cut show that a faulty condition still exists in one or both halves of the cable. With power cables this may simply be dampness running back a short way into the cable, a condition that can be cured by drying the ends out or cutting the cable back to where it is dry.

A telecommunications cable, however, may have faults at other places or may be wet for a considerable distance from the cut, perhaps back as far as the next joint. Problems on these cables due to ingress of moisture depend to a great extent on the make-up of the cable. Long wet sections will never normally be encountered on pressurized cables because there is always egress of air. Ingress of moisture can only occur at points of severe localized damage. Neither is trouble of this nature experienced with jelly-filled cables except perhaps at joints. Non-filled plastic cables, on the other hand, admit water readily but retain good insulation resistance values and maintain their performance until the water reaches a joint. The worst affected cables are those with paper insulation, be it dry or oil impregnated. Traffic is soon affected and fault location is immediately necessary.

In all cases where faulty conditions are diagnosed, further fault locations must be carried out and each one treated as a separate test, although there are the advantages that the route, route length and velocity of propagation are more exactly known than on the initial test.

5.6 Repairs

After the fault has been cut out and repair jointing is being considered, the main principle to bear in mind is not to leave more problems on the system than were there to start with! Clearly any authority or company engaged in fault location is competent to carry out jointing on its own system but, nonetheless, the following points will prove helpful to the engineer in the field who is often left to organize repairs out of office hours and in inclement weather conditions.

Temporary jointing may be necessary. The joints made should be carefully waterproofed and all joints and cable involved should be properly supported, protected and guarded. In cramped situations, the temptation to make several joints in close proximity or to subject cables to bends tighter than permitted must be avoided. The minimum number of joints should be made – after all, all joints are potential faults!

Any length of new cable let in should be the same type and size as the cable into which it is being introduced. As the existing cable may be many years old, this is not always possible. However, due regard must be paid to the choice of core metal (e.g. copper/aluminium), core insulation (e.g. paper/plastic) and core cross-section in order to obtain the best possible engineering compromise which will not result in more faults on the system at a later date. Also, any old cable recovered from construction jobs should be kept in good condition for this very purpose.

In the case of pilot and control cables it will often be necessary to let in a length of multipair plastic cable in a run of older composite cable made up of cores and pairs, the cores of which will certainly have larger cross-sections than the cores of the new multipair cable. Several pairs in the new cable should then be combined and connected to the heavier pilot/control cores, and *all* connections and core/pair identification colours and numbers recorded for the next person who has to cut into the new cable!

5.7 Reporting

Whether or not faults are properly documented and reported depends on the policy of the company or authority running the network concerned. There is an exception to this when a consultant or contractor is carrying out fault location for the body owning the network. In this case, records will probably be kept by the consultant or contractor which may or may not be passed on to that body.

Fault documentation tends to exist in one or both of two separate recording systems. The first of these is the report which should always be made in order to maintain good engineering practice. This embodies fault location log sheets such as those shown in Figs 2.1 and 2.2, suitably annotated and filed. The file may also contain basic statistical analyses such as listings of types of fault, fault incidence on different networks and voltages, etc.

The second is a formal fault reporting system which can be on a national/provincial basis or an authority/company basis. Such a system will involve the routine filling in of standard fault report forms set out with coded compartments so that all aspects of the fault location can be categorized and subsequently analysed by computer.

To repeat and augment the information given in Chapter 2, the following are some of the categories used:

- Time of occurrence
- Method of notification
- Duration of outage
- Cause/probable cause
- Age and type of cable
- Location (environment) of cable

Figure 5.2a and b shows copies of the fault report forms used by Eastern Electricity plc in the United Kingdom.

Statistics derived from these reports over a period of years provide powerful input to engineering management and help to decide policy on, for instance, types of cable to use, methods of jointing to employ, purchase of fault location equipment, training and placement of staff, aspects of communications related to notifying faults, cable laying methods, cable records, claims on third parties, etc.

158 UNDERGROUND CABLE FAULT LOCATION

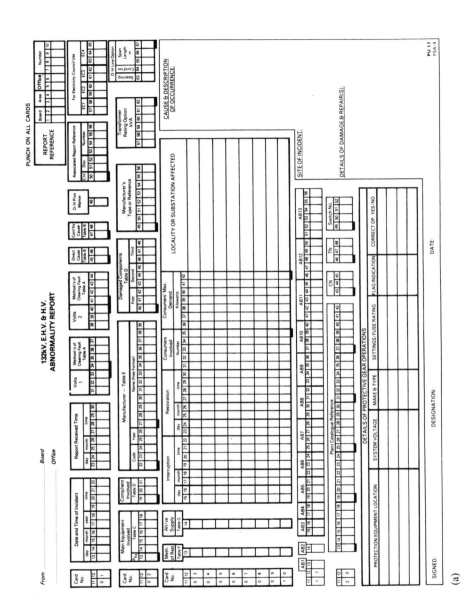

(a)

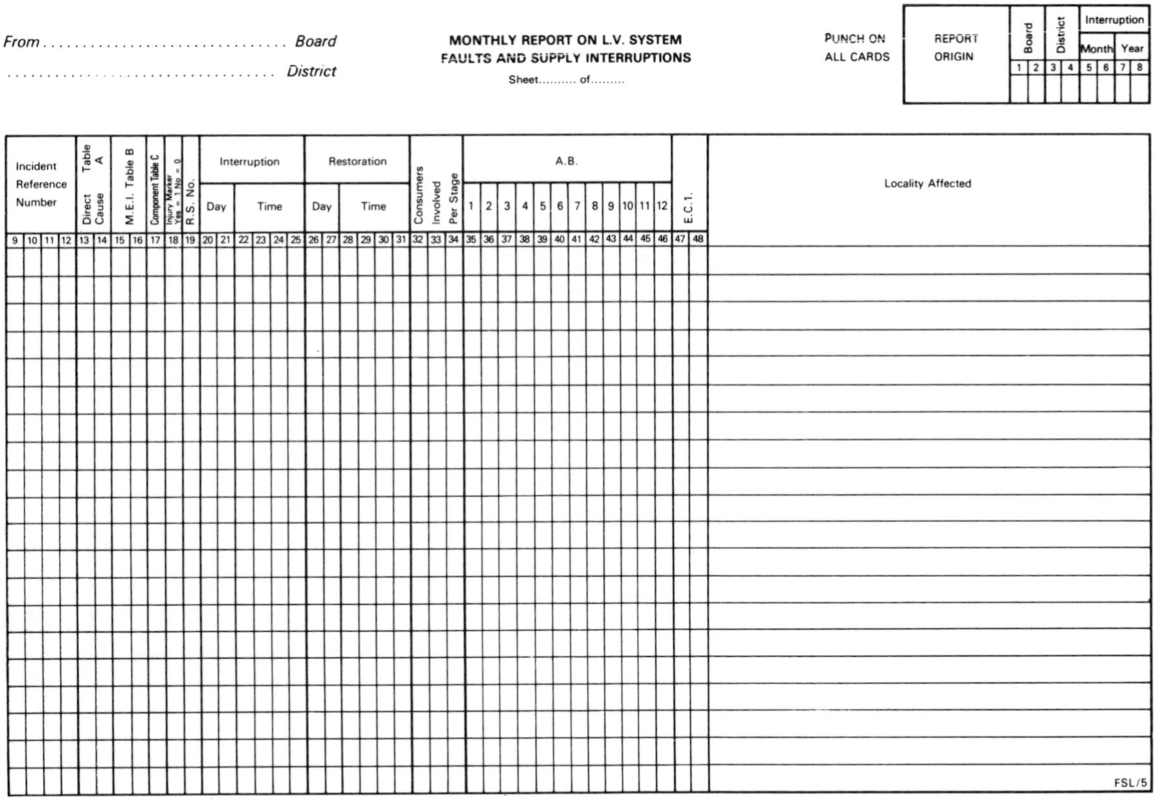

From .... Board
.... District

MONTHLY REPORT ON L.V. SYSTEM
FAULTS AND SUPPLY INTERRUPTIONS
Sheet......... of.........

PUNCH ON ALL CARDS

REPORT ORIGIN	Board		District		Interruption Month		Interruption Year	
	1	2	3	4	5	6	7	8

Incident Reference Number				Table A Direct Cause		M.E.I. Table B		Component Table C	Injury Marker Yes = 1 No = 0	R.S. No.	Interruption Day		Interruption Time				Restoration Day		Restoration Time				Consumers Involved Per Stage			A.B. 1	2	3	4	5	6	7	8	9	10	11	12	E.C.1.		Locality Affected
9	10	11	12	13	14	15	16	17	18	19	20	21	22	23	24	25	26	27	28	29	30	31	32	33	34	35	36	37	38	39	40	41	42	43	44	45	46	47	48	

FSL/5

(b)

Fig. 5.2 Fault report sheets used by Eastern Electricity plc (a) faults and supply interruptions (b) abnormality report

160 UNDERGROUND CABLE FAULT LOCATION

5.8 Notes on Part One

As its title 'general' implies, Part One covers all aspects of cable fault location, starting from basics and finishing with documentation. The information given thus far should help field engineers to become proficient in fault location generally and give them an idea of what equipment is available to match their needs.

Part Two contains the peripheral but absolutely essential subjects of route tracing, cable identification and safety, while Part Three deals more specifically with the many types of network that may be encountered.

In Part One the reader will have noted that, in this general approach to fault location, no flow chart has figured. Flow charts are a useful tool in choosing a course of action in a developing situation. However, when applied to cable fault location, a single comprehensive flow chart must assume that all modern methods and equipment are available. Patently this is not always the case. Not every field engineer has a fully equipped test vehicle on call!

It is the intention to break with tradition here and present several flow charts, each of which is dedicated to action constrained by limited kits of equipment or, in some cases, restraints imposed by the system itself. They are set out in Figs 5.3, 5.4, 5.5 and 5.6.

Figure 5.3 shows a 'best of a bad job' procedure for locations where no pin-pointing equipment is available or, indeed, where no high voltage can be used because the cable passes through hazardous areas. In this procedure it has to be accepted that the 'cut and test' solution must be used when no pre-location can be made. The careful testing and averaging from each end underlines the necessity for the elimination of errors in pre-location.

The approach shown in Fig. 5.4 is for the situation in which most operators find themselves; they have enough equipment to trace the cable, pin-point the fault and pre-locate all types of fault except the 'flashing fault'.

In Fig. 5.5 all equipment is available except that for pre-location by 'transient methods' such as impulse current and arc reflection.

Finally, the approach is given when a complete kit of equipment is available (see Fig. 5.6). Most pre-locations are made by transient methods which can also provide a good indication of the fault condition in the diagnostic stage.

162 UNDERGROUND CABLE FAULT LOCATION

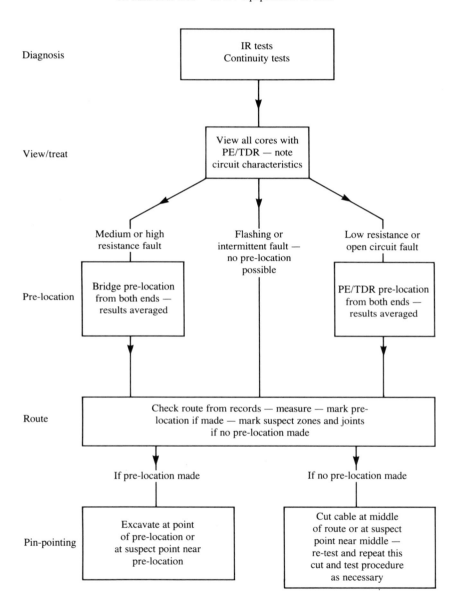

Fig. 5.3 Fault location procedure where no pin-pointing equipment is available or where high voltages cannot be used

CONFIRMATION 163

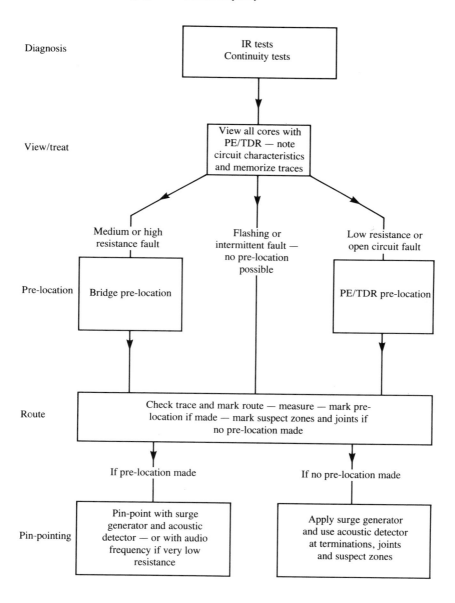

Fig. 5.4 Fault location procedure where equipment is available for all types of fault except the 'flashing fault'

164 UNDERGROUND CABLE FAULT LOCATION

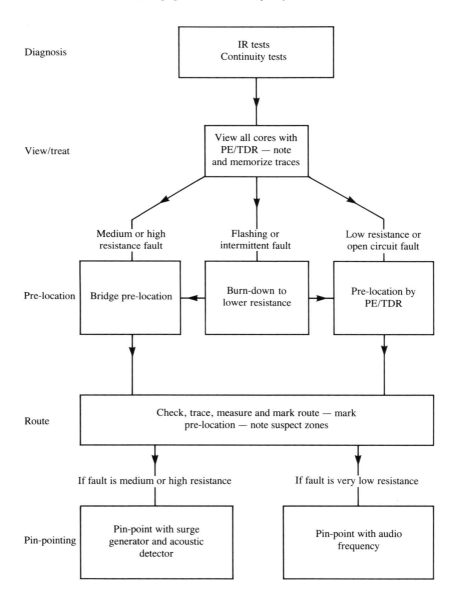

Fig. 5.5 Fault location procedure where all equipment, apart from that for pre-location by 'transient methods', is available

CONFIRMATION 165

Equipment available: IR tester, resistance bridge, PE/TDR with impulse current and arc reflection facilities, surge generator, audio frequency

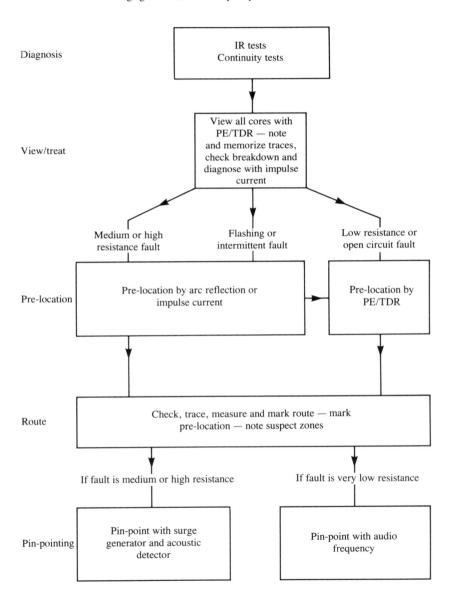

Fig. 5.6 Fault location procedure where a complete kit of equipment is available

PART TWO

Related skills and procedures

6

Route tracing

6.1 General

The tracing of metallic cables and pipes is carried out by producing an alternating magnetic field around them. This is detected overground with a search coil whose output is fed into a suitable receiver/amplifier and presented to the operator as an audible tone in headphones and/or as a deflection on a meter. Any such tone is, of course, within the audible frequency range which is why the system is normally called the 'audio frequency method'. The alternating currents in the cable or pipe, however, can be within this range or outside it. They vary from a very low frequency of approximately 10 Hz (see Chapter 10, 'control systems'), through 50 or 60 Hz mains frequencies to radio frequencies. Some of these signals are impressed on the line for the specific purpose of tracing it. Others are present anyway, e.g. mains and long wave radio frequencies. Impressed signals vary from several hundred hertz to about 200 kHz, and existing ones are 50/60 Hz and 15 to 20 kHz. Before looking at specific frequencies used in commercially available equipment, it is necessary to consider the basic principles involved in cable and pipe tracing. (For the sake of convenience, the cable or pipe may be referred to hereafter as the 'line'.)

6.2 Detection of magnetic fields

As shown in Fig. 6.1, an alternating current in a metal line produces a concentric alternating field around it. This field couples with any suitable search coil placed in it to produce a detectable alternating signal.

The illustration shows the signal pattern detected when a coil held vertically in the field is passed through it across the run of the line. A distinct minimum is noted when the coil is directly over the line because the magnetic lines of force are horizontal at that point, resulting in minimum flux linkages. As the coil is moved to either side, however, the lines of force

170 UNDERGROUND CABLE FAULT LOCATION

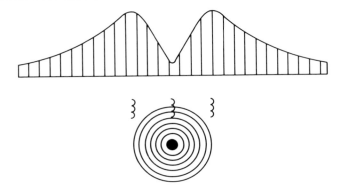

Fig. 6.1 Signal pattern across line with coil vertical

approximate to vertical and there are maximum flux linkages producing a maximum signal.

As shown in Fig. 6.2, the opposite is true for a coil held *across* the run of the line, a maximum being detected when the coil is directly over it. Also, if a coil lying above the line and at 90° to it is turned so that it is in line with the run of the line, another minimum signal is noted. This is shown in Fig. 6.3a and b. These manifestations give the operator a useful selection of options for tracing any underground metal line of significant length.

6.3 Signal sources

The matter of having a suitable signal in the line, however, is somewhat more complicated. Considering, firstly, magnetic fields that may already exist around a line, mains currents in a.c. power cables can produce massive fields. These currents are often of the order of hundreds of amperes so that, for instance, a concentric single phase cable will produce a very significant

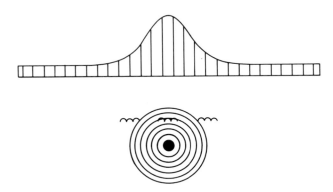

Fig. 6.2 Signal pattern across line with coil horizontal

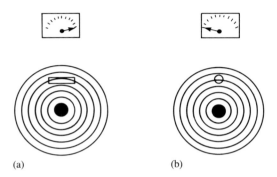

Fig. 6.3 (a) Maximum signal with horizontal coil across run of line, (b) minimum signal with horizontal coil in line with run

field. However, most systems encountered are three-phase; therefore their cables contain three or four current carrying cores. Even when single-core cables are used, they are laid either in trefoil or in close proximity flat formation.

This means that we normally have to contend with three current carrying cores within a metal sheath displaced at or about 120° one from the other. Figure 6.4a is a four-core low voltage cable cross-section and Fig.6.4b a three-core medium or high voltage cable cross-section.

In theory, if the system is perfectly balanced, the phase currents will be equal. In LV cables there will be no neutral current and the vector sum of the phase currents will be zero. In the case of three-core medium and high voltage cables this is also possible in practice but is extremely rare. Even a resultant out-of-balance current of 1 A is a significant signal source.

The low voltage cable situation is even more explicit in that there is always a resultant current to generate a traceable magnetic field. This is because, firstly, low voltage systems cannot be perfectly balanced and, secondly, because the geometry of a four-core cable means that the main

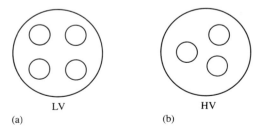

Fig. 6.4 (a) Four-core low voltage cable cross-section and (b) three-core medium or high voltage cable cross-section

current-carrying cores (the phases) are not mutually disposed at 120°. Therefore, it can always be assumed that loaded low, medium and high voltage cables can be detected and traced using 50 and 60 Hz signals.

It is most important to know, however, that live unloaded cables do not produce a magnetic field and therefore cannot be detected, identified or traced by any method involving the detection of mains frequencies.

The other circumstance in which metal lines can be traced without impressing a signal on them comes about when they carry very low frequency (VLF) signals produced by long wave radio transmissions. These occur world-wide and all underground metal lines of significant length pick up and re-radiate them. These signals are strongest on lines that are very long and have a large capacitance to earth or are well earthed at each end. Most modern cable tracing sets are equipped to take advantage of this.

Notwithstanding the existence of such 'free' sources of signal, the best signal for tracing is one that is put on to a line for the express purpose of detecting and tracing it. The signal will normally be fed into a line from a portable audio frequency generator by galvanic connection or inductive coupling, whichever best suits the situation, and it will have a frequency chosen by the operator to give the best chance of tracing the line over the longest distance. Low frequency signals (*ca.* 300 to 1200 Hz) propagate well for quite long distances. Higher frequencies (8 kHz and above) give stronger, more clearly detectable signals but they decay over a relatively short distance and are easily bled off by other metal lines connected to, or in close proximity with, the line being traced.

6.4 Methods of connecting/coupling

The best method of connecting or coupling the signal depends on one or more of many factors, among them:

- Whether a power cable is alive or dead
- Whether a telecommunications cable is clear of traffic
- Whether there is access to one or both of the ends
- Whether or how a cable is earthed
- The length of the line
- The size and make-up of the cable or pipe

The generator signal can be connected galvanically to a cable as shown in Fig. 6.5a, between any core and earth, with that core earthed at the remote end (resistive current) or between any core and earth with no short at the remote end (capacitive current). As the latter clearly depends on the capacitive reactance between the core and the sheath, it follows that a larger current is drawn at the higher frequencies. Alternatively, a high in-phase current will flow when the core is earthed at the remote end. It can

ROUTE TRACING   173

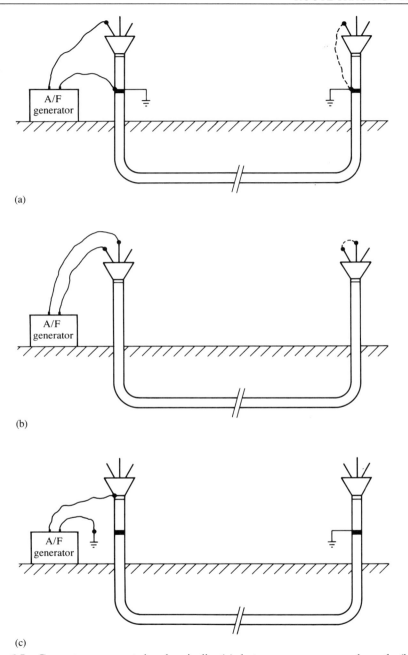

(a)

(b)

(c)

Fig. 6.5 Generator connected galvanically (a) between any core and earth (b) between two cores or (c) between the sheath and earth if the source end earth has been disconnected

also be connected between two cores, again with or without a short between them at the far end, as shown in Fig. 6.5b, or between the sheath and earth if the source end earth has been disconnected, as shown in Fig. 6.5c.

This connection is also relevant if there is no earth bond at either end. A cable is then exactly the same as a pipe; i.e. they are both metal tubes buried in the ground. If this metal tube (sheath or pipe) is unprotected by an insulator such as a plastic sheath, it is in contact with the earth at all points along its length and a resistive current will flow from it back to the transmitter ground spike, as shown in Fig. 6.6. If, on the other hand, this 'tube' is insulated, a capacitive current will flow because of the capacitance between it and the surrounding earth. The only real, practical, criterion for deciding the method of connection is *whether there is a path for current at the frequency chosen*.

Inductive coupling is carried out in one of two ways. The first uses an aerial coil built in or fixed on to the transmitter. The second uses 'transmitter tongs' fed from the transmitter and clamped around the cable or pipe. This latter method is extremely effective in the field as the tongs can be clamped around almost any dead, live or loaded cable or any pipe which they are big enough to fit round. The only coupling not recommended is with a loaded single-phase a.c. cable. In normal use the tongs simply comprise a transformer with a multiturn primary, inducing a current in the cable or pipe which is a single-turn secondary.

There is no problem with even very large three-phase cables carrying heavy currents as the resultant out-of-balance current is quite small. However, the tongs around a single-core cable carrying perhaps hundreds of amperes are trying to perform as a current transformer with a single-turn primary and dangerous voltages can occur at the transmitter output connections which can cause damage to the equipment or to personnel.

Transmitter tongs vary in size, the largest being able to encompass a

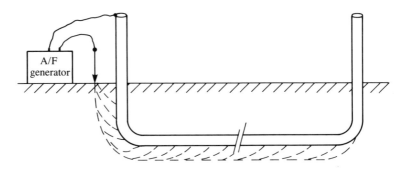

Fig. 6.6 Signal injected between metal line and earth

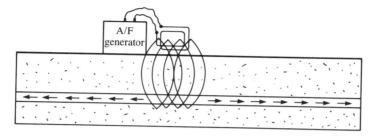

Fig. 6.7 Induced signal using aerial coil

cable or pipe of about 100 mm diameter. This is sufficient for most cables but only for the smaller range of pipes. They have a split ferrous core and the ends of the two 'C' shaped halves of this core are ground to a good fit to allow no air gap when closed. Great care must be taken to keep dirt, grit and corrosion off these ends as even a small air gap ruins the magnetic circuit and stops signal propagation.

To induce a signal into a buried line using the aerial coil, the transmitter is simply placed directly over it and in line with it as shown in Fig. 6.7. The field generated by the aerial coil couples with any metal line within its ambit and induces a signal into it. The frequencies used here are *ca.* 8 kHz and above, the lower ones being much less effective for inductive coupling. If the aerial is at an angle to or to one side of the line, coupling is worse and less signal is induced. It should also be borne in mind that metal lines other than the known one can have a signal induced in them if they also run close enough to the transmitter aerial.

Depending upon the frequency chosen, the line can be followed confidently for up to several hundred metres using frequencies in the 8 to 12 kHz range, but often only for less than 100 metres with the higher frequencies. This is no major problem, however, as the transmitter can be repeatedly moved up the route from one certain location to another.

With the minor exception already explained, the transmitter tongs can be placed around any accessible metal line. This can be a power or telecommunications cable in a pillar, cabinet, service termination, joint hole, etc., or a water or gas pipe wherever exposed. There simply needs to be enough room to clamp the tongs around the line.

Figure 6.8 shows transmitter tongs clamped around a cable earthed at both ends as is normally the case. The single-turn secondary is the closed loop made up of the cable and the earth return path. It is also clear that the tongs must not be clamped around the cable *above* the earth bond because there would be *no secondary circuit*. A gross example of this same mistake is shown in Fig. 6.9 in which the tongs are clamped around a cable with no earth bond at either end.

176 UNDERGROUND CABLE FAULT LOCATION

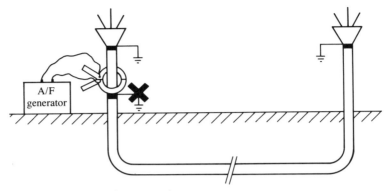

Fig. 6.8 Induced signal using transmitter tongs

The question 'Has the current anywhere to go?' is much more relevant here than in the case of galvanic connection where there is often some sort of capacitive load. When coupling with the tongs, there must always be a closed circuit loop as the secondary. Only then will current flow. The lower the resistance of the loop the better.

6.5 Practical approach

There are many facets to finding and tracing services in the field, so this very practical matter will be covered under the following specific headings:

- Receiver aspects – tracing, marking and depth checking
- Transmitter aspects – hints and sound practice
- Ground survey
- Metal detection
- Joint location

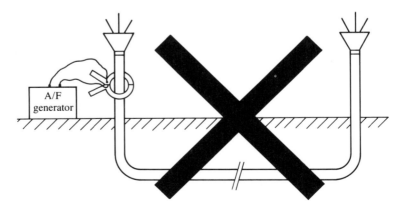

Fig. 6.9 Incorrect use of transmitter tongs

6.6 Receiver aspects – tracing, marking and depth checking

The first point to make about receivers is to emphasize that there are a great many available! There are receivers with separate search coils. Some are combined in one unit. Tone output may be via a loudspeaker or headphones or either. Visual indication can be by meter or LCD bar indicator with or without alphanumerics. Some receiver sets employ one search coil, others use several. Some have a direct depth readout, some do not ... and so on and so forth.

This discourse refers to any of these features and facilities as necessary, while keeping as its main premise the basics involved in tracing and depth checking. The procedures to be described apply equally to signals of any frequency being transmitted into the line. Any particular receiver will, of course, be equipped to deal with these and will have switched selection of the frequencies available from its own transmitter, e.g. 50/60 Hz (or aperiodic for low frequencies), 1, 8/10/12, 33, 100 kHz, etc.

The receiver frequency should be selected to that being transmitted and a check made in the vicinity of the line at a known position near the source end to ensure that the signal is present. If this signal is at or below about 1 kHz, the signal heard will be that received. However, for all higher frequencies, the tone heard will not be at the transmitted signal frequency as this will be near the limit of human hearing or above it. Therefore a tone usually of about 1 kHz is produced electronically which is proportional to the received signal.

The receiver will have amplitude/sensitivity controls. While still at the known location of the line, these controls should be set so that any maximum reads approximately $\frac{3}{4}$ f.s.d. on a meter (if applicable), and any minimum reads near zero. Audible levels should be set for comfortable recognition in whatever ambient noise obtains.

There is a very important point regarding headphones. While high quality, ambient noise protected headphones obviously give very good reception, they are dangerous as they block out extraneous sounds such as those from traffic and nearby cranes and earth moving machinery. Much safer are the cheaper, 'stethoscope' type headphones or a small loudspeaker built into the receiver. With these, the operator hears a perfectly adequate signal and is still aware of potential dangers in the vicinity.

The operator is now ready to walk and trace the route. By far the best indication of the run of the line is the minimum detected with the coil(s) vertical. The operator should repair to a point several metres to one side of the probable route and, holding the search coil vertical and still, walk quickly across the line. The signal detected will rise slowly to a crescendo just before the line is reached, drop suddenly to a marked minimum over it and quickly rise again before finally tailing off. The exact minimum point

178 UNDERGROUND CABLE FAULT LOCATION

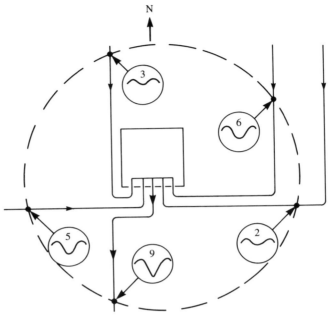

Fig. 6.10 Multiple signal paths

should then be carefully located and marked. The maximum indication with the coil at 90° to the run can then be checked, as well as the other minimum noted when the coil is in line with the run. Levels can be readjusted as necessary.

It is often the case that there are other lines in the area that also carry signals. Such a situation is shown in Fig. 6.10.

Let it be assumed that a signal of, say, 10 kHz has been produced in one of the cables shown by coupling with the transmitter tongs. In this example this is the line with the heavy outgoing arrows on it which depict the current being injected. This current in the 'wanted' cable must return to source via *any other convenient path*. The other cables out of the substation certainly qualify as other paths, as do any other known or unknown pipes or cables running through the area. This means that there are several other lines in the vicinity of the substation with detectable signals on them. There are four such routes in the example shown.

Although this presents a problem, the very multiplicity of return paths helps to solve it. Kirchhoff's current law states that the sum of currents entering or leaving any point is zero. Quite simply, in practice, if we call the outgoing current I, then the currents returning on the other four cables must each be of the order of $I/4$, depending on the quality of each path and the existence or otherwise of more paths. Therefore the signal level on any 'unwanted' line is normally lower than that on the one selected. The exception to this is when there is one line in and one line out of a particular location. In such cases, each line can be traced until it is obvious which one is going towards the correct destination.

The next action is quickly to circumnavigate the substation to check all the signals with regard to signal form, strength and the depth of the line. Firstly the coil should be placed across the run of one of the lines, in this case say the west-bound one, and the sensitivity adjusted to give a mid scale reading. In all, five locations will be noted, but the sharpness of the minimum indications will differ. By this time the operator will have a good idea of which route is the one to be traced because the minimum over it is very sharp. It then remains to set up the levels again if necessary so that no maximum indication goes over scale, and to check each location, very carefully this time, for the form or steepness of the minimum and the levels of the maxima as scale readings. In the illustration 'south' is the wanted line as the readings are:

- West, 5
- North, 3
- First east/north, 6
- Second east/north, 2
- South, 9

There is one circumstance in which the above readings can be misleading: when one line is much deeper or much shallower than the others so that it gives a smaller or larger reading than if it were at the same depth as them. This is exactly why depth checking should be carried out as a routine matter. If the equipment has the capability of measuring the value of the signal *current* in the line, this will confirm the route being sought. As a final check, the operator usually knows where the remote end of the wanted cable is. In extremely difficult situations it can be traced back from the far end.

Having decided which line is the correct one, the operator can now follow the route. This should be done using a continuous transmitter signal and holding the search coil vertical but moving it from side to side of the route. If the pace is kept brisk, this gives an excellent impression of the minimum over the line and the shoulder signals on each side. It is much more difficult to 'keep in touch' with this pattern if the pace is slow and the search coil is poked at the ground with small random movements.

If the signal suddenly diminishes or disappears, the operator should make a mental note of the place at which this happened and keep on walking for another few metres. Should the signal not reappear, it is

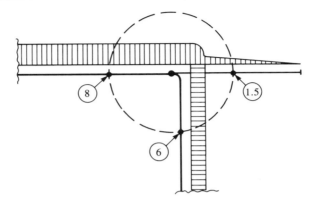

Fig. 6.11 Correct tracing procedure at branches

probable that there is a bend or tee joint at that spot. The operator should then hold the search coil in a steady vertical position and rapidly walk around the spot in a circle of about 5 to 10 m diameter, if obstacles allow it. In the example shown in Fig. 6.11, the line will be detected again at two other points, the continuing main and the tee branch.

The relative strengths and patterns of the signals can then be established and each leg traced back to interpolate the position of the tee joint. In Fig. 6.11, the scale readings obtained with the coil across the line are:

- On the main before the joint, 8
- On the main after the joint, 6
- On the tee branch, 1.5

Tracing can then be continued on the main route or the tee branch as desired.

A bend or turn in the cable or pipe can be confirmed in the same way. In this simpler case, the ongoing signal path will quickly be established at the same signal strength. A useful hint when tracking bends is to orientate the search coil in line with the route before it turns, move the coil slowly along the supposed route and monitor its orientation for minimum readings. As the line turns, the angle at which the coil gives a minimum changes, showing that the run of the line is also changing. This is particularly useful when trying to establish whether or not a cable or pipe is turning into a building at a certain point.

The route should be marked every 2 to 10 m depending on later requirements to check any particular stretch for any reason, e.g. road works or cable fault pin-pointing. It is utterly fallacious to think that anything but the shortest of routes can be effectively memorized. Routes should always

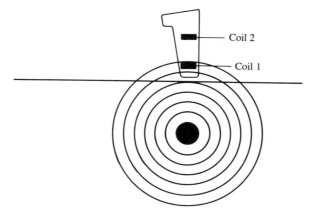

Fig. 6.12 Two-coil search array for depth checking

be marked. The depth can also be marked as required at points of interest or as a check on marking accuracy (see the following comments on depth checking).

All tracing sets are capable of checking depth. Those equipped with a two-coil search array (usually encapsulated in the hand-held receiver unit) simply require a 'depth' button to be pressed after the instrument has been confirmed as being over the line. Figure 6.12 shows clearly that, with such an array, the lower coil, labelled coil 1, is closer to the magnetic field source and therefore is cut by more dense lines of force than the upper one, coil 2. As this difference is due to depth, the actual depth is simply calculated and displayed by the receiver circuitry. As a matter of interest, such a double-coil assembly gives sharper indications and is less susceptible to interference fields than a single coil when in tracing mode. The coils are connected differentially so that the interference field (which is more or less of the same strength and sense in each) is cancelled out.

Depth checking with a single coil is not automatic but simple to carry out. The actual route position is first carefully marked with the coil vertical (minimum indication). With the coil reorientated at 45°, it is moved slowly *at ground level*, to each side in turn. As shown in Fig. 6.13, a minimum is detected on each side as the coil takes up a position perpendicular to the lines of force. This occurs on a line running down to the cable or pipe at 45° or, in other words, when the coil is pointing to it. Marks should be made on the ground at each of these two minimum positions.

If the magnetic field is symmetrical about a vertical axis, the distances L_1 and L_2 will be the same and will each be the same as the depth, D (for circular fields). In practice, the lines of force are flattened slightly and most

182 UNDERGROUND CABLE FAULT LOCATION

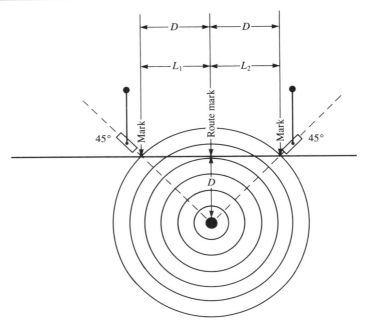

Fig. 6.13 Depth checking with a single coil

depth checks have errors of up to 10 per cent, the line being slightly deeper than indicated. (Double-coil systems give an error of about 5 per cent.)

Where the magnetic field is more grossly deformed by the presence of other metal lines, a properly executed depth check not only gives the depth but shows up errors in route marking. Figure 6.14 shows a misshapen, more or less elliptical field surrounding a line. When the above test is carried out, the distance L_1 from the initial route marking is much less than the distance L_2. The geometry of the situation dictates that, in all cases, the depth is $(L_1 + L_2)/2$.

The fact that L_1 and L_2 differ, however, is a clear indication that the first route marking was wrong. These marks were made at points of minimum indication, and quite correctly so, as the lines of force are indeed horizontal at these points. The depth check proved, though, that the field was deformed at certain points, thus enabling a correct route marking to be carried out.

6.7 Transmitter aspects – hints and sound practice

The various methods of putting a signal on to a line, either by direct galvanic connection or inductive coupling, have been mentioned already, but more detail is required.

The matter of frequency is vital. The lower frequencies of several hundred

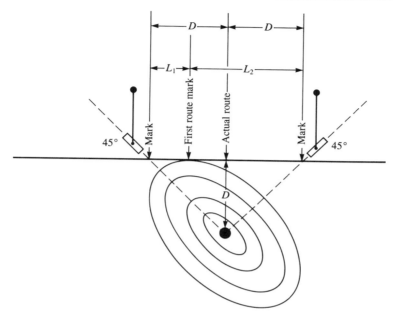

Fig. 6.14 Effect of misshapen field

to around one kilohertz are suitable for tracing the longer lines as any capacitive signal bleed-off is small. Therefore the signal travels farther and is not picked up too easily by other (unwanted) lines. They are also best in situations where a good low resistance path can be established. This can often be accomplished quite easily using relatively short overground connections. For instance, Fig. 6.15 shows how a signal can be injected into a short, very low resistance loop by connecting to pipes or cable armour/sheath in adjacent houses. This will pick out the two service runs and part of the main (shown shaded) with absolute certainty. Such a connection is much better than a direct or inductive coupling at a termination.

There are a few coupling tongs available operating at about 1 kHz but, generally speaking, this type of transmission is best carried out using

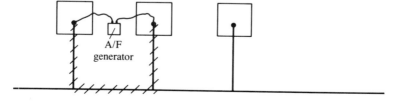

Fig. 6.15 Easy tracing along short loop

frequencies in the region of 8 to 30 kHz. The use of transmitter tongs is by far the most popular and effective method, being convenient, reasonably selective and applicable to almost every metal line, live or dead, which is accessible at one point, be it a termination or in-line position. Furthermore, the signal propagates along the whole length of the line being traced, which forms the outgoing leg of a closed resistive circuit, and the signal current in it is constant along its length up to any splitting-off point (see Fig. 6.10).

The matter of how far a signal can be expected to propagate along a given line is a vexed question in field operations. The terms 'significant length' and 'significant distance' have already been used in this section. The basic fact is that a current will flow along the whole length of a resistive loop but signals into a capacitive load will diminish as the capacity decreases with length traversed. In addition, it falls off even more due to bleeding-away on other paths.

The rate of fall-off cannot be easily estimated or forecast. It depends very much on frequency. Signals at 30 kHz and above may propagate for distances of from fifty to several hundred metres, whereas one at 10 kHz may be traceable from several hundred to one or two kilometres. With a connection between core and sheath at one end of a dead cable at this sort of frequency (i.e. using the core/sheath capacitance), it can well be possible to successfully trace a route for several kilometres, the signal only falling off seriously in the last 100 m or so. A check on the remote section of the route can then possibly be made from the far end. If little success is obtained, it is useful to try bunching cores together as one connection. This is often a good ploy with telecommunications cables.

The best results are obtained by trying several combinations of frequency and connection/coupling and choosing whichever gives the best signal.

The method of inductive coupling using the built-in (or added-on) aerial coil presents no distance problems. When the transmitter has been correctly placed over the line at a known location, as already described, the matching (or method selection) switch is set and the power turned up. The receiver and search coil should then be taken about 10 m away from the transmitter before tracing is commenced. Otherwise, the search coil can receive signals direct from the transmitter and not the underground line. Initial tracing will be easy because signal levels will be high. They soon diminish, however, and sensitivity will have to be increased to compensate for this. It is a mistake to continue too far like this because serious signal drop-off will be evident after about fifty to several hundred metres and attempts to trace further will expose the operator to the risk of wandering off on a false line which has picked up some signal. It is far better to stop while the signal levels leave no cause for doubt and bring the transmitter up to a new position that has been very carefully checked and marked. Obviously, this method is easier with two people. Once the transmitter has been set down in its position, a further section can be traced. This process can be repeated as often as is necessary.

One further method of connection is possible with some equipment. This employs a blocking filter which allows a galvanic connection to be made via the filter to live low voltage mains. The filter blocks the mains voltage, so protecting personnel and equipment, while allowing the signal to pass into the live mains.

Some transmitters are portable battery/mains units. The batteries are chargeable and will deliver a power of up to around 20 VA for 4 to 6 hours on an overnight charge. This is sufficient for a day's work as the use is rarely continuous. When used from mains, the output power is increased typically to about 50 VA.

Another group of low power consumption transmitters relies on large dry batteries or banks of dry cells which have a life of 20 to 40 hours depending on the particular unit.

6.8 Ground survey

It is a very common requirement, not just to trace a known line but to survey a particular area and locate all the services within it. These include:

- High and medium voltage electric cables
- Low voltage cables
- Telecommunications, signal and control cables
- Water pipes
- Gas pipes
- Buried metal objects
- Non-metallic pipes and culverts

No matter what total area is involved, it is best initially to mark out an area of about 20 m × 20 m.

Apart from the tracing and detection equipment, different sets of pegs or stakes and marking chalks or paint should be to hand for marking out the different services. A plan of the area should also be available showing as many of the (supposed) service routes as possible. This is not to be trusted but, as the various lines are ascertained, they can be marked on it and checked with whatever services are shown in order to build up a composite picture. When the survey is complete, this plan should be annotated and dimensioned so that a properly drawn-up plan can be made later.

Having marked out the area, the operator should first traverse it both 'up and down' and 'across' with the equipment set to detect 50/60 Hz signal. This will usually result in a line or several lines being established which can immediately be designated as loaded electricity cables. (Just occasionally, a significant mains hum is detected emanating from a very large iron pipe which is carrying earth return currents.) The route(s) should then be marked within the 20 m square area. Any one of these lines could be a cable or a bunch of cables.

The next phase involves carrying out an identical check but, this time, with the equipment switched to the 'radio frequency' mode, i.e. set to pick up re-radiated very low frequency (VLF) signals. This will locate most or all of the other metallic lines within the area. These in turn should be marked. The need for different types or colours of pegs should now be apparent. It is essential to mark everything located in a totally unambiguous way.

Depending on the equipment in use, there is another way of generally checking a small area of 20 m square. A search coil and receiver are located in the centre of the area with one operator, and set up with roughly 'mid range' visual and audible indication levels. In some equipment the coil is a special one which can be left standing in one position. Other sets utilize the normal search coil used for tracing, but it must be orientated in the maximum signal position and kept facing the other operator who is controlling the transmitter. This operator sets the transmitter up in aerial mode and moves it slowly around the periphery, always pointing to the receiver in the centre, as shown in Fig. 6.16, which depicts a normal coil being rotated with the transmitter movement.

Every time the transmitter passes over a buried line a peak signal will be

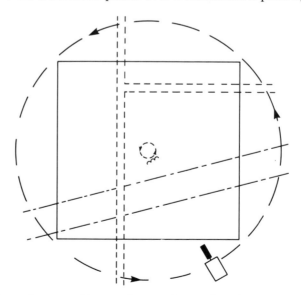

Fig. 6.16 Ground survey with normal transmitter and receiver

ROUTE TRACING 187

noted at the receiver and the position of the transmitter is then marked. Of course, not all the lines pass under the receiver coil; therefore these peaks will vary greatly in magnitude. Sensitivity adjustments are made to suit. The main thing is to mark the transmitter position and move on to the next peak as a detailed examination will be carried out later.

Once the transmitter has travelled full circle and the marks have been made, it is placed over one of them and the operator with the receiver carries out a normal trace of that line, following it up to and beyond the marked-out area and marking it carefully. It is good practice then to move the transmitter to a point exactly over the newly marked line and trace the line *back* through the area and beyond as necessary. The next phase of the operation is the tracing of all of the lines out of the area in a similar way and the marking of route and depth. Other 20 m square areas can then be marked out and surveyed as necessary. In describing how to survey an area in a general way not specific to any particular manufacturer's equipment and methods, it is difficult to fully discuss particular circumstances, such as identifying lines in close proximity and discriminating between them. There are equipments available with 'built-in intelligence' which are capable of indicating the *direction* of the tracing signal current and measuring its value in milliamperes.

To further assist the beleaguered engineer, in every practical situation there are factors that can be turned to advantage. There will be some house, substation, exchange (central office), pillar, compound, pole with cable, trial hole, etc., giving access to a cable or pipe so that a connection or coupling can be made with it. This particular line can then be traced back through the survey area and positively identified. Such clues, together

with those gleaned from personnel, plans and ground markings, enable the astute engineer to reach the correct conclusion.

There remain two types of buried object that have not as yet been discussed: non-metallic pipes, ducts and culverts, and masses of metal. The former do not come within the scope of this work. Tracing is usually carried out by following signal emitting sondes which are introduced up the pipes. The latter can be located by the methods given in Sec. 6.9, 'metal detection'.

The most common type of buried metal object (not metal line) is the manhole, handhole, link box or valve cover that has been overgrown, covered with soil or by road surfacing. Clearly, when one of these is found and dug out, it can be identified as belonging to a telecommunications, electricity or water line and thus that line can be properly identified for what it is.

All the information gained regarding route, depth, type of service, etc., should be meticulously recorded on the rough plan for subsequent drawing up before any pegs are removed or marks obliterated.

6.9 Metal detection

Some route tracing equipments have the capability to locate buried metal. Other special instruments are available that are designed for this. Either way, the equipments function in one of two ways:

1 The search coil forms part of an oscillator circuit which the user sets to a null or very low frequency tone when the search coil is positioned over a metal free area. When it is moved over a significant mass of metal the inductance of the coil is changed and the frequency of the tone emitted also changes. In the specially designed instruments the coil is separate whereas, in general purpose ones, it is usually wound within the transmitter housing. In the latter, the metal detection function is a useful extra facility but the same degree of accuracy and depth penetration cannot be achieved as with the dedicated units.

2 The transmitter and receiver are capable of being mounted on a boom at 90° one to the other, as shown in Fig. 6.17. A fine mechanical adjustment allows the mutual alignment of the transmitter and the receiver to be altered until a null signal is achieved. When the assembly passes over a piece of metal or a metal line, the signal peaks.

The most effective metal locator is a dedicated unit operating on principle 1.

When adjusted for a null or low frequency tone, the pitch rises as the coil

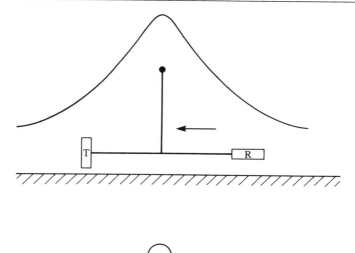

Fig. 6.17 Transmitter and receiver mounted on a boom at 90° to one another

is passed over the edge of a buried cover. If the cover is large there can be a minimum at its centre and another maximum at the far edge.

The maximum depth at which a cover can be detected depends on its size and is approximately 1 m for large covers about $\frac{2}{3}$ m across. Small covers of less than 20 cm cannot be found if deeper than around $\frac{1}{3}$ m. It is not possible to be exact in these values but, generally speaking, covers are not very deep and can be located without too much trouble, particularly if they are known to exist in a certain area.

Before starting any search, the unit should be set up over a piece of ground with no buried metal present. Then, if possible, it should be checked for response and levels over a visible metal object.

Equipments operating on principle 2 are route tracing sets which usually operate at around 30 to 100 kHz, mostly in the inductive mode. They are quite often made up of plastic transmitter and receiver housings which fit together as a base and lid. Their survey mode, whereby the transmitter and receiver are mounted on a boom as shown in Fig. 6.17, is most effective down to depths of a little over 1 m for the location of cables and pipes as well as metal. This method is also good for discriminating between lines that are close together and give one peak response at the first pass. A reduction in sensitivity and renulling then produces separate peaks for each individual line.

6.10 Joint location

There are two audio frequency methods of joint location, the first and most effective being a variation of the twist method of fault location already described in Chapter 4. As shown in Fig. 6.18, an audio frequency signal is fed into two cores of a cable which have been shorted together at the remote end.

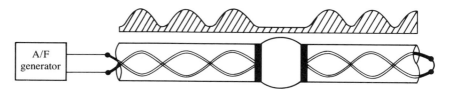

Fig. 6.18 Joint location by twist method

Before and after the joint, the normal rising and falling signal pattern is detected but over the joint there is a lull. Depending on the orientation of the search coil, this can also be a longer peak, but the important point is that there is a most marked change at that point. This is due to the fact that the regular twisting of the cores is interrupted because they are spread apart and straightened for ferruling within the joint.

When carrying out this test it is essential to cover the route at a steady but relatively fast walking pace so that the smooth rise and fall of signal before the joint is firmly in the mind as a pattern at a certain constant frequency. This is suddenly disrupted as the operator passes over the joint and the difference is so marked that there can be no doubt about the presence of the joint. This is especially so if the operator continues over the joint at the same pace and immediately picks up the same pattern on the far side of the joint. If done properly as stated, even a change of cable size or type can be detected as the rise and fall will be more rapid on one side of the joint than on the other.

As stated before in the description of the twist method, there is only one way to be sure of covering the route at a fast pace. That is to trace and mark the route beforehand.

The second audio frequency approach is as shown in Fig. 6.19. If the sheath is accessible in the vicinity of a suspected joint position, a signal of 8 to 12 kHz is injected between the sheath and a ground spike. When the route is traversed with the receiver and 'A' frame fitted with capacitor plates, a distinct signal peak is detected over the joint. This is because the joint has more contact points with the earth than the cable itself and the currents leaving it are more dense in the region of the joint.

Should the suspected joint position be somewhere in the middle of a long

ROUTE TRACING 191

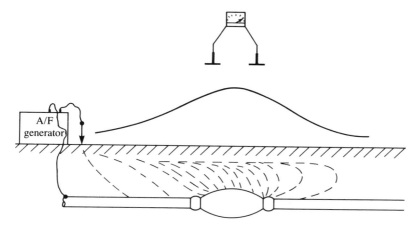

Fig. 6.19 Joint location using 'A' frame

route, the same effect can sometimes be achieved using the transmitter aerial coupling method with the transmitter placed over the cable 20 or 30 m away from the area the joint is thought to be in. If there is no success, the transmitter should be moved, just in case it has been placed directly over the joint.

7

Cable identification

7.1 General

Engineers in the field regularly have to face the problem of identifying one particular cable either in isolation or from a bunch of cables in a hole, trench or cable rack. If a telecommunications cable is cut by mistake, there can be severe disruption of vital traffic (sometimes security related). The wrong identification of a power cable, however, can have catastrophic or even fatal results if the cable is cut into.

The only way to avoid such serious trouble is to adopt a 100 per cent certain procedure or set of actions that *cannot lead to an accident*. There are methods that help to identify a cable and a few that positively do so. Even with the latter, a cable should never be cut with a saw on the evidence of an identification test alone. Instead, the procedure should be:

1 Apply the identification method correctly.
2 Confirm by physical means and prove the cable is dead and is the correct one or, in the case of telecommunications cables, prove it is the correct one.

Taking point 2 first, having decided that the cable which has been identified is *almost* certainly the right one, there are secondary actions that should be taken. They are:

1 For a telecommunications cable, strip the sheath and identify the cores carrying signal (see later).
2 For a power cable (low voltage), strip the armour and sheath and test the cores to prove the cable is dead and then check electrically that it is the correct one.
3 For a power cable (medium and high voltage), 'spike' the cable and then identify electrically.

CABLE IDENTIFICATION 193

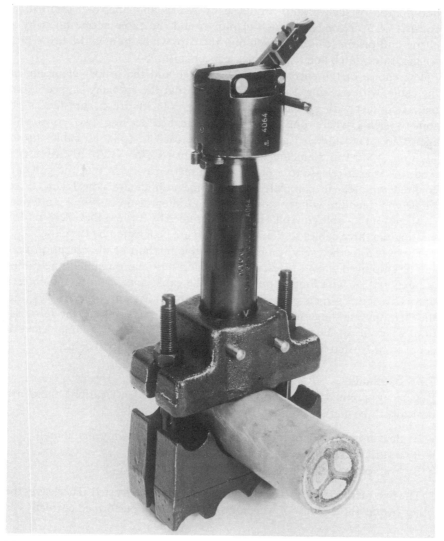

Fig. 7.1 Cable gun spike (*source:* Accles and Shelvoke Ltd, Birmingham, UK)

EXPLANATORY NOTE

A cable spike, as used some years ago, was an earthed chisel which was driven into the cable with a hammer. Though this is infinitely better than sawing through the cable, there are obvious dangers if a live cable is spiked by hand in this manner.

The only totally safe cable spike is a cable gun spike. Figure 7.1 is a photograph of such a gun, produced by Accles and Shelvoke Limited. It consists of a heavy base which clamps round the cable surmounted by a barrel and piston (chisel), breech and the cap, which houses the firing pin, trigger, safety latch and trigger release mechanism.

The choice of the correctly rated cartridge and the proper clamping of the cable, with packing if necessary, ensure that the specially shaped chisel makes contact with at least two conductors. (There are circumstances on some systems where one core can be earthed leaving two cores still energized.) The standard gun can safely be used to spike all cables up to 113 mm (4.5 in) in diameter and carrying up to 275 kV. A lanyard is run from the trigger release to a point several metres away so that the gun can be fired remotely in complete safety. Some utilities are now laying three single-core medium voltage cables bound at intervals in trefoil. An alternative base for the cable gun spike has been designed by Eastern Electricity plc for holding the three cables so that at least two are penetrated by the chisel.

There is a further very useful and positive method of identification that can be employed when the test point is near a termination and all the cables are exposed. A wire loop is put aroung the cable *thought to be* the correct one. This loop is then passed along the cable, past any other cables and through any building entry ducts until it arrives physically at the termination, where it can be confirmed that it is around the correct cable. The eye is not to be trusted in tracing out any one cable in a group of cables.

7.2 Methods

Regarding point 1 of the procedure for cable identification, there are basically three methods of identifying a cable. They are:

- By detecting and quantifying the audio frequency tracing signal on it
- By a variation of the audio frequency 'twist method'
- By low voltage d.c. impulse

The first method is often correct but should never be trusted. This leaves the other two methods, both of which are trustworthy if performed correctly.

TWIST METHOD
The method simply requires the connection of an audio frequency generator to two cores of the cable which are shorted together at the far end, as in joint location. A small coil, specially designed for this purpose, is placed on the outside of the cable at the point of access at which it needs to be identified. If this coil is passed along the surface of the cable, the typical rising and falling 'twist method' signal pattern will be detected, provided that a stretch of cable has been exposed which is long enough to account for several half-twists of the cores. For telecommunications cables this

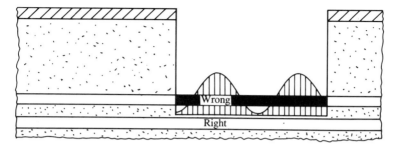

Fig. 7.2 A dangerous situation when identifying cable by twist method

requires only a short length, but a large power cable may need to be exposed for over a metre of its length.

It would seem that detecting this rise and fall pattern should suffice to identify the cable, because all the other cables in the vicinity will have no signal or a steady lower signal on them with no peaks. There is, however, one very dangerous circumstance in which this is not true. Figure 7.2 shows a cable exposed on the bottom of a hole over which a good rising and falling signal has been detected.

In this particular case, the *correct* cable (which is producing the signal) is directly under it, buried just below the floor of the hole! There is only one positive way of being certain in this and all other situations; the small coil is passed *around* the cable. If it is indeed the correct cable, two maxima and two minima will be detected, as shown in Fig. 7.3.

These maxima and minima are slightly asymmetrically disposed around the circumference because no two cores of a three-core cable can lie on a

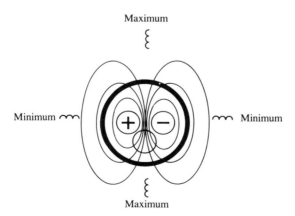

Fig. 7.3 Cable identification by twist method – disposition of maxima and minima around cable

diameter. Once these maxima and minima are detected, the cable will have been properly identified. Small telecommunications and control cables can be identified in exactly the same way, although there is more asymmetry in the disposition of the maxima and minima as the pair selected may be central or peripheral.

Another way of identifying telecommunications cables and cores or pairs is to use audio frequency equipment, the receiver of which can be fed by coupling tongs and a small inductive probe (basically a small search coil in the bottom of a pencil-like housing). The wanted cable is detected by clamping the receiver tongs around each cable in turn and choosing the one with the largest signal. Once the cores are exposed, the bunch of pairs or group, if known, is divided into two and each checked with the probe. The half giving the best response is further subdivided until the wanted core or pair is finally identified. In this final check, the side of the probe at the end should be offered to the conductor, not the tip, as this gives the maximum response from the coil inside the probe.

LV D.C. IMPULSE

As its name implies, this method has nothing to do with the audio frequency methods. It is a test that relies totally on the detection of signal *direction* and it can be used on any type of cable. This, of course, means that it is the only method that can be used on single-core and concentric cables because there is no twist effect from the cores of such cables that would allow use of the twist method.

The equipment used comprises a generator, fed from the mains supply because of power requirements, which produces a series of steep fronted d.c. current surges. These are high current surges of between 60 and 300 A, depending of the manufacturer, and they occur every few seconds.

The receiver is basically a centre zero galvanometer whose input comes from a pair of receiver tongs placed around the cable being checked. The galvanometer scale plate is coloured green to the right-hand side of the zero mid point and red to the left of it. This signifies that a deflection into the green zone is of correct polarity and indicates that the cable clamped *is* the correct one. A deflection to the left into the red zone means that the cable clamped *is not* the correct one.

The receiver tongs cannot give any output if the cable through them carries a steady d.c. current. However, the heavy current pulse generated in the cable has a fast rate of rise which produces flux linkages and an output voltage in the secondary winding feeding the galvanometer. This manifests itself as the needle 'kicking' to one side or the other and then returning to zero. The indication is quite positive as the magnitude of the deflection does not matter, only the direction.

CABLE IDENTIFICATION 197

Therefore, providing the proper procedure is carried out, this test will identify the correct cable in the presence of others. There is one aspect of this procedure that must be checked and double checked. The receiver tongs have to be clamped around the cable in the *correct direction*. To this end, they are marked with a distinctive arrow which must always point away from the source when they are clamped around a cable. If they are clamped on the wrong way, the correct cable will be taken as a wrong cable and any wrong cable as the right one. This is clearly of the utmost importance. Figure 7.4 shows how the d.c. pulse generator is connected and what indications are obtained with the tongs correctly clamped around the correct cable (top) and two incorrect ones (middle and bottom).

From this it can be seen that the generator output is connected to a core of the cable to be identified and this same core connected to earth at the remote end. The pulse travels along this core and back to the generator via *any earth path*, as indicated by the small arrows. Therefore there will be various smaller values of current flowing in all other cable sheaths but, if the tongs are clamped around any of these, the galvanometer will kick to the red (wrong) zone. Amplitudes will vary and are not relevant.

The correct cable, on the other hand, has two currents in it – the large outgoing current in the core and the smaller return current in the sheath. The outgoing current predominates and produces a kick to the green (correct) zone, thus identifying this cable as the correct one.

There is a circumstance in which this system does *not* work. That is where the *only* return path is the sheath of the correct cable and the go and return currents cancel out. Such a situation is, however, most unusual and, in any case, if this is known or suspected, connections can be made at the remote end to create other return paths.

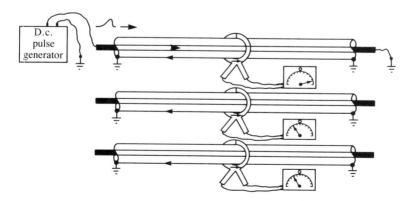

Fig. 7.4 Cable identification by d.c. impulse method

7.3 Identifying live cables

An extremely important point must be made here about determining whether a cable is *live* or *dead*. If the cable is carrying current, an external magnetic field will be produced and this can be detected. However, it often occurs that a cable, particularly an MV/HV cable, is off-load, i.e. switched in at one end and open at the other end. It carries no current and therefore does not produce an external magnetic field.

If such a cable has a metal sheath, there is no way of detecting the electric field from outside the cable. This field stretches from the core(s) to the sheath and ends there. The only way to detect it is to break into or remove the sheath.

There did exist a method, in use about forty years ago, whereby a pellet of the lead sheath was carefully removed so as not to damage the outside belting papers, and a probe inserted to detect the electric field. Not only was it less trustworthy than the method about to be described, but it was only used on known MV/HV cables as an expedient to avoid damaging the cable by spiking prior to making tee joints. This, of course, flies in the face of safe practice and the method has been discontinued for a long time. In any case, it was never used to discriminate between MV/HV and LV cables.

This is the very area that requires attention. Deciding whether an exposed cable is an MV/HV or LV one is a vital duty imposed on the operational engineer *daily*, because LV cables are opened and jointed live on a routine basis in all distribution authorities and companies.

Very strict safety procedures are followed to make sure that the cable to be opened up is the correct low voltage one. Cable records are carefully checked and the jointer is then normally issued with a jointing card showing the type and size of the cable, and its location and disposition are indicated by a dimensioned sketch. Tracing and identification methods and instruments are also employed as necessary.

Despite all these precautions, there is still a risk that cables thought to be low voltage will be opened up and found to be medium or high voltage. Although most modern cables of different voltages are different also in construction and aspect, there remain buried in the ground, in close proximity, many thousands of kilometres of MV/HV and LV cables that are identical in external appearance.

In recent years this problem has been addressed by a UK Regional Electricity Company. An instrument named VODCA (voltage discriminator cables) has been developed, tested and put into operation by Yorkshire Electricity. Figure 7.5 shows the unit, manufactured under licence from Yorkshire Electricity by M and B Systems Limited, which is a carefully engineered voltage sensor set up to give positive unambiguous, failsafe discrimination between MV/HV and LV cables.

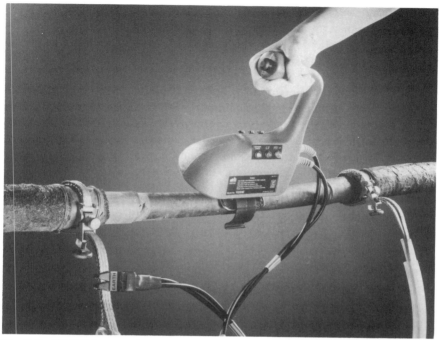

Fig. 7.5 VODCA, manufactured under licence from Yorkshire Electricity (*source:* M and B Systems Ltd, Stalybridge, UK)

VODCA requires the removal of the metallic sheath in order to detect the electric field. It is accepted that this in itself constitutes a hazard if the cable turns out to be HV or MV, so its use must be made additional to and following all other precautions and identifying methods. It is important that the use of VODCA must not in itself constitute an additional risk. The instrument therefore has a high degree of insulation for the operator and a specially contoured sensing plate in order not to create additional stress on the cable belting papers.

As already mentioned, there will always be a chance of a MV or HV cable being accidentally opened up. There are on record a number of occasions where this has occurred and a potentially serious accident averted by the use of VODCA.

In keeping with normal safe jointing procedures, the procedure is as follows. Armouring and protection are removed to expose the metal sheath. The bonding strap is fitted spanning the section of sheath to be removed and the correct length of sheath is then marked out and carefully eased off so as not to damage the belting insulation. The instrument earth lead is then connected to the bonding strap which switches it on. The sensing plate is placed on the exposed cable insulation and the test button pressed. The test button applies an internally generated voltage to the sensing plate. This tests the total functioning of the instrument *in situ*. On releasing this button the illumination of the red LED indicates an MV/HV cable and the green one an LV cable.

To ensure that the maximum field is measured it will be necessary to rotate the instrument through a minimum of 90° in case, initially, the sensor plate is lying between two cores or over a neutral core. However, if at any time a red LED indication is noted, the cable must be MV/HV. Green indicates an LV cable. If there is no indication at all, the cable is dead or there are hidden metallic screens blocking the electric field.

Even if there is no red indication, there is continued observance of strict safe procedures and belting insulation is carefully removed until the cable is confirmed as LV. If evidence of screens or three-core construction is noted, stripping is abandoned.

8

Electrical safety

8.1 Introduction

Of all the aspects of electrical engineering, safety is the most important. The adoption of safe procedures is even more critical in the field of fault location because fault location is an extra, specialist activity grafted on to the day-to-day work of system operations, maintenance and construction. Furthermore, it is carried out by an engineer who might well be a part of the normal maintenance team, but who is just as likely to be an 'outsider' from another department or even another company.

This greatly increases the chances of accidents happening through ignorance, misunderstanding or poor communications. The following comments are limited to safety aspects associated with fault location on electrical systems.

8.2 General

Accidents do not just happen. They are made to happen by a series of apparently unconnected facts, circumstances and shortcomings which, when they all come together at the wrong time, almost guarantee trouble! Among these are:

- Poor regulations or lack of regulations
- Poor training or lack of training
- Poor discipline
- Bad management
- Bad communications
- Notices and labels
- Poor equipment or lack of equipment
- State of health
- State of mind

- Weather
- Distractions
- Assumptions

We can do something about most of these, such as creating and adhering to sound regulations and systems of control and operation, but the way to prevent accidents is to build in basic safeguards and checks which serve to break any chain of circumstances that would otherwise lead to an accident.

To repeat and augment the brief statement on safety in Chapter 2, the ground rules are:

1. Instructions given must be clear and preferably in writing.
2. Comments and instructions received must not be believed but must be checked no matter from whom.
3. Assumptions must never be made.
4. No action should ever be taken on a *signal*, prearranged or otherwise, or after a prearranged time lapse.
5. An engineer should carry out all operations with or with the knowledge of another person.
6. During any sequence of actions, the operator should intentionally *stop, think* and *check* before carrying out any operation.
7. Conductors must be tested with a proven tester and then earthed with the correct earthing equipment where applicable (HV) before being touched.
8. All cables and HV test equipment should be fully discharged after testing, even if automatic discharge is built into the equipment.
9. Above all, written safety rules and procedures should be strictly adhered to at all times.

8.3 Safety regulations

Every organization which owns an electrical network has formal safety regulations for operations on that network. In these, the matter of testing for maintenance or fault location purposes is covered in detail and correct procedures are laid down for getting access to a high voltage cable. They require:

- A switching programme for the isolation of the cable
- The locking off with special locks of all circuit breakers and switches involved
- The posting of 'caution people working' notices on equipment being worked on
- The posting of 'danger' notices on adjacent live equipment
- The proper earthing of the cable at all its extremities
- The issue of a sanction for test to the person who is to be in charge of the testing.

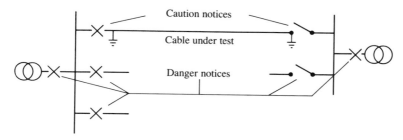

Fig. 8.1 Location of earths and statutory notices

Figure 8.1 shows the location of earths and statutory notices on equipment associated with the cable to be tested.

Earths are applied by using the feeder circuit breakers or isolators themselves if they have a 'feeder earth' position, or with special portable earth fittings provided by the manufacturer for this purpose. In all other cases, e.g. on some cubicle equipment and overhead lines, flexible earths are connected with operating rods after the conductors have been proved dead.

Every authority has its own form of statutory 'permit to work' and 'sanction for test'; examples of these, from Eastern Electricity plc, UK, are shown as Fig. 8.2a and b. After a 'sanction for test' has been issued, the earths can be removed for testing. They are reapplied when all tests are complete and the 'sanction for test' is cancelled.

All areas in which work is to be carried out are cordoned off, usually with special coloured plastic fencing posts and chain or tape. This is particularly important in outdoor switchyards with row upon row of identical circuit structures. Again, danger notices are posted on all adjacent live equipment.

Even in indoor situations with metal-clad switchgear, dangerous situations can arise if notices are not properly posted and correct test procedures are not carried out. For instance, test access may be required to the cable termination shown in Fig. 8.3. It is extremely dangerous (and always totally forbidden) to insert the hand or any metal object up the orifice at A before a tester has been applied, when the approved test extension can be inserted.

Potentially more dangerous is touching the cable termination at B after removal of the current transformer chamber cover. This is at the rear of the switchgear, often out of sight of the front, and the absence of danger notices could mean the cover is taken off the wrong circuit. Also, the fact that the circuit breaker (CB) is open and isolated does not mean that the cable is dead. It could be energized from the remote end.

On MV and HV systems, there is some danger of explosion and fire, but the main danger is of electrocution.

204 UNDERGROUND CABLE FAULT LOCATION

| CONTROL Ref. No. | **EASTERN ELECTRICITY**
PERMIT-TO-WORK | SF 5031/JUL90
049651 |

1. ISSUE To: ....

The following High Voltage Apparatus has been made safe in accordance with the Distribution Safety Rules for the work detailed on this Permit-to-Work to proceed:-

....

....TREAT ALL OTHER APPARATUS AS LIVE

Circuit Main Earths are applied at : ....

other precautions and information required to be entered by Distribution Safety Rule 3.2.1(b), 4.6.2(c), 5.5.3 and 5.10.2(b) and Approvals: ....

The following work is to be carried out:- ....

DIAGRAM

Signed .... Time .... Date ....

2. RECEIPT I accept responsibility for carrying out the work on the Apparatus detailed on this Permit-to-Work and no attempt will be made by me or by persons under my charge to work on any other Apparatus.

Signed .... Time .... Date ....

(a)

Fig. 8.2 Typical forms (a) permit to work, (b) sanction for test

ELECTRICAL SAFETY 205

| CONTROL Ref. No. | **EASTERN ELECTRICITY SANCTION-FOR-TEST** | SF 5045/JUL90 (SF 158) 07051 |

1. ISSUE To: ....
The following High Voltage Apparatus has been made safe in accordance with the Distribution Safety Rules for the testing described on this Sanction-for-Test to proceed:-

The points of isolation are:

....
....
....
                                                    ....

Circuit Main Earths are applied at:
                                                    ....
....
....
....

other precautions and information required to be entered by Distribution Safety Rule 3.2.1(b) and Approvals: ....
....

Brief description of testing to be carried out: ....
....

DIAGRAM

Signed .... TIME .... DATE ....

2. RECEIPT   I accept responsibility for the testing described on this Sanction-for-Test and for taking the precautions necessary to prevent danger.

Signed .... Time .... Date ....

(b)

206 UNDERGROUND CABLE FAULT LOCATION

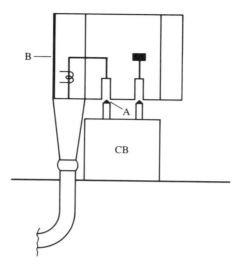

Fig. 8.3 Typical switchgear layout at cable termination

Low voltage systems are totally different and, paradoxically, can be more dangerous. This is because, although safe working practices are well documented, there is not the same formal approach as on HV systems. Work is carried out daily on live conductors using rubber mats, gloves and boots. Also a shock is rarely fatal and the operator can quite wrongly become overfamiliar and lax. The main dangers are from explosion and burns. Whereas HV systems are protected by very fast acting protection relays, LV systems are protected by fuses. Very high fault levels exist, particularly near substations. A short circuit on the LV side of an MV/LV transformer produces an awesome cloud of ionized gases and molten copper as the protection on the MV side of the transformer can be very slow to clear a fault on the LV side.

Low voltage cables should always be tested with proven test lamps (not neon) or a voltmeter before being touched and the utmost care should be taken when handling tools, clips and pieces of wire in the vicinity of LV distribution boards.

PART THREE

Specialized areas

9

Power cables

9.1 Faults on low voltage systems

GENERAL

If a faulty cable is simply terminated at both ends, the fault can be diagnosed and located by one or more of the methods already described, no matter what the length or nominal voltage of the cable.

The simple fact that a cable may be a low voltage one is no problem; it is the corollary that most low voltage cables are *permanently loaded* that creates difficulties which are occasionally insurmountable.

In this discussion on low voltage cable fault location it is assumed that a cable has loads which cannot all be disconnected. Therefore the normal method of diagnosis, the IR test, is meaningless. The meter will simply show the IR value of the bulked loads and the fault in parallel.

TYPES OF SYSTEM AND TYPES OF FAULT

Most low voltage systems fall into the categories of industrial or domestic supply and street lighting. Single-phase, three-phase four-wire and two-phase three-wire underground residential distribution (URD) cables are in common use with or without metal sheaths and running at a.c. voltages between 100 and 400 V plus.

Street lighting circuits are not considered here as the lamps can often be disconnected and there are many other particular circumstances relating to fault location on such networks. They are therefore treated separately in Chapter 12.

Most low voltage cables in industrial environments can be isolated at each end and carry no tees. Thus they do not present any special problems, other than the obvious limitation on test voltage.

Domestic supply systems in many countries, however, consist of three-phase cables with solid tee joints providing services to the premises. These services may be single- or three-phase and the service terminations may be inside the premises and therefore normally inaccessible, or external and therefore accessible for disconnection.

The very worst systems, from the point of view of fault location, are those, such as the ones found in the United Kingdom, that have multiple single-phase services into inaccessible premises. Most of these systems were (and still are in some areas) laid with four-core copper or aluminium cables with a metallic sheath and steel wire or tape overall, the neutral being earthed at the source substation only.

Faults on such systems can be:

- Core to sheath, permanent or transitory*
- Core to core, permanent or transitory*
- Open circuit core, with or without paths to other cores or earth
- Open circuit neutral,† with or without phase involvement

Another type of system in common use currently is a multiple earth system. These systems are normally laid with three-core cable with a metal sheath which serves as neutral and earth combined. This sheath/neutral is earthed at source and also connected to earth spikes at various points along the route (usually at joints) and again at the very end of the cable, such that there is an earth on each side of any service. The earth resistance at each of these spikes should be less than 10 ohms.

This means that a low resistance system earth can be achieved conveniently which can be a distinct advantage in areas where, perhaps due to rocky ground, it is difficult to achieve a low value with one earth rod at source. Much more significant, however, is the cost saving in using three-core instead of four-core cables, and many kilometres of such cable have been laid in the last twenty or thirty years.

Once multiple earthing was sanctioned all networks on which it was to be used were designated *multiple earthed* and run as such. This is then the normal type of low voltage system on which the poor fault location

---

* A permanent fault will be evidenced by a fuse having ruptured initially and by the fact that fuses inserted subsequently rupture immediately. Transitory faults, however, manifest themselves through reports of 'flickering lights' or a ruptured fuse. In the latter case, a fuse inserted subsequently will often remain intact and maintain supplies for a short or, indeed, a long time afterwards (see Figs. 1.7a and b and the associated text).

† A broken neutral on a four-wire system with single-point earthing at source produces phase to neutral voltage imbalance after the break, the only voltage reference for the neutral being the connected loads. This phenomenon is covered fully in the section on 'diagnostic problems' on page 211.

engineer is required to operate. It is a complete hybrid, with some four-core copper, some four-core aluminium and a lot of modern three-core cable with sheath/neutral earths at many points. Because of these multiple sheath earths none of the sheath fault location methods can be used. Even on single-point earth systems, the sheath is usually in contact with the ground at many points such as metal box joints and places where the outer protective serving of the cable is damaged.

A very different type of low voltage system exists, however, on which ground contact methods of fault location can be used to great effect. These are the systems laid with plastic insulated cables with *no metal sheath*, such as triplexed or flat formation URD cables. The outer plastic sheath is very often broken or burnt away, producing contact between a core and earth. Even if pre-location is difficult, pin-pointing is readily carried out using a ground contact method (see Chapter 4, 'pin-pointing'). These cables are not covered in this discussion on difficult permanently loaded networks.

DIAGNOSTIC PROBLEMS

As mentioned at the beginning of this section, normal diagnostic procedures cannot be used because the system loads shunt the fault. Diagnosis, therefore, falls into two broad categories:

- Evidential diagnosis and
- Diagnosis by re-energizing

The first simply means the use of common sense in interpreting factual evidence and fault symptoms *as found*, i.e. before any attempt is made to re-energize the system. Paradoxically, the very complexity of a loaded low voltage system can produce evidence that indicates the fault position or leads to a location – another example of the fault 'giving itself up'.

212 UNDERGROUND CABLE FAULT LOCATION

Examples of this are:

1 One consumer only is off supply, all fuses in the substation being intact. The fault is obviously an open circuit in the service joint or along the service cable (see the section on 'methods available' on page 217).
2 All consumers are off supply beyond a certain point on the route and all fuses in the substation are intact. This indicates that a phase is open circuited on the main cable.
3 Consumers after a certain point have damaged appliances due to high voltage, all substation fuses remaining intact. The neutral is broken and the consumers lying between the break and the end (or a second neutral break on a multiple earth system) are subjected to low or high voltages depending only on the connected loads at any given time. The neutral is floating. This is shown in Fig.9.1.

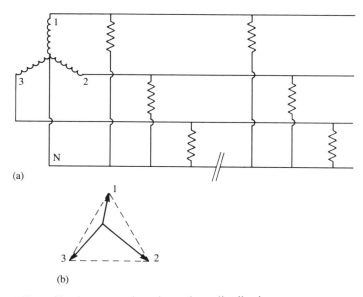

Fig. 9.1 Effect of broken neutral on three phase distribution system

Another common type of 'self-diagnosing' fault is the permanent core-to-core fault. This is diagnosed simply by checking the voltages on the outgoing cable stems with filament test lamps or a voltmeter. A cursory test from neutral to each outgoing phase can show that each phase is live. However, it is essential that, on all occasions, tests are also made between phases and across each fuse.

It is often the case that a phase-to-phase fault has occurred and one fuse only has blown. If, for instance, cores 1 and 2 are welded together at the fault and phase 1 fuse has blown, all cores will be alive and all consumers may well be on supply, some being fed through the core-to-core fault. All single-phase equipment will function. Three-phase equipment will not.

In this situation a complete test (such as should *always* be carried out) will indicate:

1 – N   phase voltage
2 – N   phase voltage
3 – N   phase voltage
1 – 2   no voltage*
2 – 3   phase-to-phase voltage
3 – 1   phase-to-phase voltage

At this juncture it is little short of a criminal act to replace fuse 1!

The fact that cores 1 and 2 are so well burnt together as to maintain the flow of load current indicates that a low resistance core-to-core fault exists which can be found by the twist method already described.

In all other circumstances, diagnosis has to be carried out by *re-energizing*. If the operator finds a fuse ruptured, the phase concerned has to be re-energized, not only to try to restore supplies but to check if the fault still exists.

It is also possible, on many occasions, to 'treat' the fault by repeated re-energizing and thus produce a burn-off to create an open circuit situation or the burning together of two cores, both of which configurations lend themselves to location. Furthermore, any change caused at the point of fault will facilitate location by PE/TDR, as described in the section on 'sound approach' at the end of Sec 9.1.

HISTORY

Before the advent of PE/TDR testing on low voltage systems in the sixties, low voltage faults were located by the audio frequency twist method, if core-to-core, or by repeated energizing to change the fault condition, or by removing links to sectionalize the system, or by listening at suspect points when re-energizing, or by a method known as 'cut and test', which was, is and always will be in daily use around the world. It consists of cutting the cable at or near its mid point and identifying the faulty half by re-energizing.

---

* This means that, on the cable, cores 1 and 2 are at the same phase potential and a test across the phase 1 fuse holder will indicate phase-to-phase voltage, thus confirming that cores 1 and 2 are at the same potential and the phase 1 fuse has ruptured.

The faulty half is then cut in half again and the faulty 'quarter' identified by re-energizing, etc. Thus, in three or four cuts the fault can be found. It has distinct advantages:

1 It is certain.
2 It is reasonably quick.
3 The consumers see something being done.
4 It allows the reconnection of sections proved healthy and the progressive restoration of supplies.
5 It requires no test equipment.
6 People are working who might otherwise be waiting around.

However, it involves the digging of many unnecessary holes and puts many extra joints on the system which devalue the cable and are themselves potential faults for the future.

Burn-down using system voltage was also attempted, with varying degrees of success, by re-energizing via a current limiter made up of electric fire elements or a water resistance. Many expensive fuses were expended on a routine basis in re-energizing cables to check if faults still existed, to restore supplies or to modify the fault condition. To save this gross wastage of fuses two approaches were tried: prospective fault current indicators and rewireable fuse arrangements or circuit breakers for re-energizing.

The former enjoyed only limited success, as they all employed some form of limiting resistance or choke and, if the fault happened to have 'gone away' at the instant of connection, load current in the limiter simply produced a volt drop across it, thus resulting in the application of too low a voltage to the cable/fault and a 'no fault' indication.

POWER CABLES 215

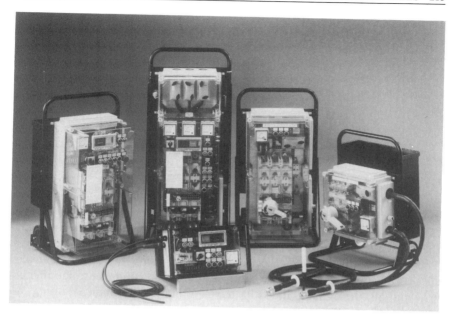

Fig. 9.2 The FRED family (*source:* Campbell York Ltd, Guisely, UK)

The latter usually incorporated old style rewireable fuse housings, second-hand d.c. circuit breakers and improvised connections, all of such dubious construction that this type of rig was incapable of being officially specified for use by any authority. The fact that such devices were in daily use by expert operators was, however, clearly indicative of the basis fact that some such device was necessary for the hard pressed operator in the field whose only recourse was to re-energize a cable with, perhaps, a very close-up fault on it giving fault levels in the region of 25 MVA!

The need for a safe and acceptable version of these devices was addressed by the author in the period 1966–1969. A fault re-energizing device was produced with a fully rated air circuit breaker, fuse protection and properly engineered connectors. This was accepted by and used on the networks of the Eastern Electricity Board in the United Kingdom.

In the early seventies it was developed further and became known as FRED (fault re-energizing device). It was put out to industry to be manufactured commercially and this unit and its derivatives have been in constant use ever since. FRED's family has grown to include three- and single-phase auto reclosing versions (FREDA series), as well as the manual versions (FREDY series). Figure 9.2 is a photograph of the range currently manufactured by Campbell York Limited.

216 UNDERGROUND CABLE FAULT LOCATION

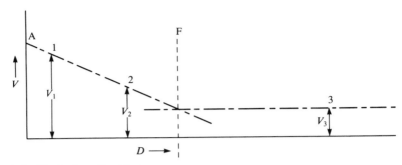

Fig. 9.3 Typical gradient test

Rating is obviously a problem in that such equipments need a permanent rating of several hundred amperes in order to feed normal cable loads and the capacity safely to break huge fault currents of thousands of amperes. They successfully do this. Electronic versions using thyristors were also successfully developed by some cable fault location equipment manufacturers, but all these had, and still have, a limited permanent rating of, typically, about 200 A.

Following tests in the late sixties using a method called 'Transgradient', a promising development was made in the early seventies by the UK Electricity Council Research Centre. This was an 'instantaneous' potential gradient method which was subsequently licensed for further development and manufacture by Location Techniques Limited and sold under the name 'LOTEC'.

Potential gradient methods had been in use for many years, particularly on street lighting cables. Figure 9.3 shows a typical gradient test where a current is passed through the fault and voltage readings are taken at several points of access down the line. Beyond the fault the voltage is low and constant, depending on the fault arc voltage (in practice the graph after the fault can rise slightly if the reference conductor is carrying return current). The approximate location of the fault is, however, indicated quite clearly where the 'gradient to fault' line intersects the 'beyond the fault' line. The problem is that loads are permanently connected, access points are few and normal low voltage faults are transitory. Voltage readings therefore have to be taken by several measuring devices spread along the cable *when the fault arcs over*.

After early tests with the transgradient method which used capacitor storage, the LOTEC system became available in the mid seventies; this utilized three or four transient recorders triggered by voltage depression and stored several voltage cycles encapsulating the disturbance (as in Fig. 1.7a). An interrogator unit was then used to measure the voltage recorded

POWER CABLES 217

by each transient recorder unit at exactly the same point in time. These values, say V_1, V_2 and V_3 in Fig. 9.3, were then used in a simple formula, or plotted out, to give a location.

This system had a degree of success in the late seventies and early eighties but fell into disuse because of the advent of more sophisticated transient techniques and the ever-present practical considerations such as:

- Difficulty in getting suitably spaced access points
- Availability of dedicated or trained staff
- Correct interrogation of the recorder units in their correct order
- Interpretation of 'no trigger' occurrences

METHODS AVAILABLE

Audio frequency

It is surely accepted that, because of the complexities of fault location on multi-branched and loaded low voltage systems, pre-location is never easy and often impossible. The direct overground audio frequency methods, therefore, are most powerful tools in the fault location engineer's armoury, providing they are fully understood and only applied in the very particular circumstances in which each is most effective. These methods are:

- The twist method, single- or double-coil
- The double-coil method for core-to-sheath faults
- The field turbidity method for core-to-sheath faults
- The 'A' frame ground contact method for core-to-earth faults where there is no sheath

All these methods are fully described in Chapter 4, 'pin-pointing', but there are some factors peculiar to low voltage fault location that need to be mentioned.

General

A low voltage cable is rarely much more than a few hundred metres long. The route has to be walked anyway, and it is always good practice to trace the cable (see the section 'sound approach' on page 223). Trying an audio frequency method is relatively easy and does not take up too much time.

Twist method

Frequencies of 1 to 10 kHz should be used, never the lower ones, so that as little signal as possible passes through the house loads and energy meters.

The cable should always be traced first so that, as mentioned earlier in Chapter 4, the test can be carried out at a fast walking pace.

When connecting the transmitter to a low voltage core-to-core fault, there is an aspect of diagnosis about matching, as the best match gives a rough idea of the fault resistance and thus indicates whether the test is likely or unlikely to be successful. If the system has a separate neutral, it is worth disconnecting this from earth to reduce signal current return paths.

It is always better to use a double-coil assembly, if available. The coils should be spaced one 'half-twist' apart. A much clearer rise and fall signal pattern is obtained in this way.

Double-coil method for core-to-sheath faults

This is rarely successful on low voltage cable faults because of the multiplicity of joints which produce changes in the magnetic fields. However, it should be borne in mind as 'worth a try' in circumstances such as those where a fault lies on a relatively uncluttered cable section, for instance between a distribution pillar and a link box or cut.

If the audio frequency generator has been connected between the faulty core and sheath in order just to trace the cable, as is usually the case, then it is convenient to try this method and get a quick yes or no result.

Field turbidity method for core-to-sheath faults

Likewise, this method is even easier to employ once the generator is connected to the faulty core and earth for tracing. While conducting the trace itself, a weather eye should be kept open for short stretches where the 'valley' of the maximum – zero – maximum signal pattern across the cable seems less deep and steep sided (see Fig. 4.20a). If detected, such a section can then be traversed carefully to examine the minimum signal axis pattern shown in Fig. 4.20b.

Any promising stretch should then be marked on the ground and on plans. Even if the operator is unsure of these indications, they will provide a definite confirmation of the fault position if they are in the vicinity of a location obtained by another method.

Finally, if one of these suspect zones happens to be over the site of a recent excavation, the fault will almost certainly have been found, and, moreover, in a very short space of time, by the application of sound procedures and common sense!

'A' frame ground contact method for core-to-earth faults

As mentioned at the end of the section on 'types of system and types of fault', ground contact methods are not normally applicable to fault location on complex loaded systems with earthed sheaths/neutrals. However, there are occasions when they can be used, such as when a fault lies on a relatively short length of unloaded cable, say between two cuts, and a core or the sheath is in contact with the ground at the fault.

The audio frequency 'pool of potential' methods set out in Chapter 4 include detailed accounts of 'A' frame fault tracing.

PE/TDR

The most important fact to bear in mind when using PE/TDR techniques on complex low voltage networks is that discrete features, even those from good shorts and breaks, cannot normally be recognized in among the multitude of features making up a typical low voltage PE/TDR trace. Figure 3.34 gives some indication of how the many reflections and re-reflections make all but the early part of the trace almost meaningless.

On the other hand, a gross mismatch (short or open) will be clearly visible, even on the most complex of cables, if it is very close to the point at which the PE set/TDR is connected. This may come about by accident or design. If the fault is a good short circuit, then luck is needed. The fault may be fortuitously near to the point of connection or the operator may have connected the set at several extremities or access points and luckily come across an indicative trace.

In the case of breaks, however, the situation is very different. The approximate position of a break tends to be known due to knowledge of which consumers are off supply. The set can thus be connected at an access point near the break and a good open circuit fault feature will be seen.

This is particularly so when only one consumer is off supply – the so-called 'service fault'. It should be noted here that many PE sets/TDRs have been damaged in the past by the a.c. mains voltage appearing at their terminals when connected like this. Several manufacturers have addressed this problem and produced special cost effective short range sets with built-in protection against mains voltage. Indeed, some are designed to function while connected to live mains. Examples of these are the TELEFAULT P240 shown in Fig. 3.37e, the short range service cable fault locator TELEFAULT SRS and the T510 shown in Fig. 3.37b.

There is a conflict of interests regarding the best pulse widths to use on low voltage systems. On the one hand, a very narrow pulse produces very sharp trace features, giving resolution well within one metre, but the pulse is seriously attenuated after only a short running distance. On the other hand, a broad pulse will not be attenuated very much, but neither will it produce sharp trace features, and discrete features are difficult or impossible to identify.

The trade-off to be sought is: for the recognition of discrete features, use a narrow pulse on short ranges. For more general use on short and medium length cables use an instrument with a broad pulse and facilities for easy comparison of healthy/faulty or before/after traces, such as one of those in the TELEFAULT range. Further advantages incorporated in this range are: hardware and software filtering and completely isolated outputs for differential connections to the healthy core (or balancing unit) and the faulty core to give suppression of unwanted common reflections.

In the absence of very specialized low voltage fault location equipment, however, the soundest approach to PE/TDR testing on complex low voltage systems is the intelligent use of units with memory for comparison of healthy and faulty cores or, more particularly, the faulty core before and after the fault has been modified, usually by re-energizing.

Figure 9.4 shows traces recorded on a very complex low voltage network in North London of all three phases both before and after the rupture of a 600 A fuse inserted in phase 1. The set of 'before' traces cannot be analysed with any degree of certainty, but the fault position is quite clearly indicated in the second set. The original faulty phase, phase 1, has been blown apart and phases 2 and 3 have assumed a very low resistance to earth.

It is essential that the faulty core trace is put into memory before anything else is done, so that any change thereafter may produce a difference. Care should be taken to set up the instrument in exactly the same way with respect to range, pulse width, amplitude, balance, etc., when viewing subsequent traces, so that like is compared with like. The more expensive and sophisticated sets cater for this automatically. If a simple set is in use, the settings must be noted down or marked.

When all services are three-phase, small differences between the faulty phase and a healthy one may be indicative, but this is not so when the services are single-phase. Then the traces will differ anyway and only a gross difference should be taken as having been caused by the fault.

By far the most successful and surest approach is to try to modify the fault condition by re-energizing the cable, preferably with a safe and convenient device such as a FRED. This has the same basic human appeal as the arc reflection technique in high voltage fault location; that is to say, any difference that appears subsequent to the initial memorizing of the trace *must* be the result of the operator's action and must therefore mark the fault position.

POWER CABLES 221

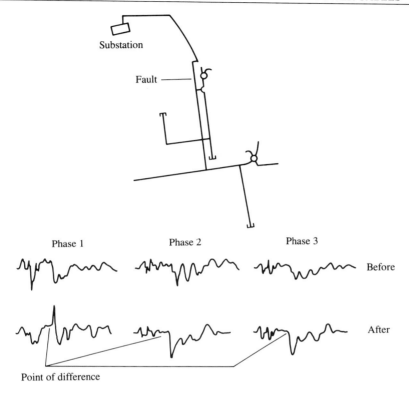

Fig. 9.4 Traces recorded on a very complex low voltage network before and after the fuse rupture

Transient methods

This same philosophy applies to modern transient methods except that, once the cable has been re-energized, traces representing the healthy and faulty states are memorized and superimposed automatically. The point of divergence of the two traces indicates the fault position.

Basically, apart from the LOTEC voltage gradient system mentioned in the section on 'history', there is just one transient method that can be used on low voltage faults, i.e. the application of PE/TDR while the fault is arcing over. However, not all PE sets/TDRs are designed for this. It is necessary to use a dedicated low voltage fault location instrument such as the T276 from Biccotest Limited, shown in Fig. 9.5, or the TELEFAULT P240 from Hathaway Instruments Limited, shown in Fig. 3.37e. The T276 injects a train of low voltage pulses into the re-energized faulty cable via a filter which protects the signal source from damage but still allows the pulses through.

222 UNDERGROUND CABLE FAULT LOCATION

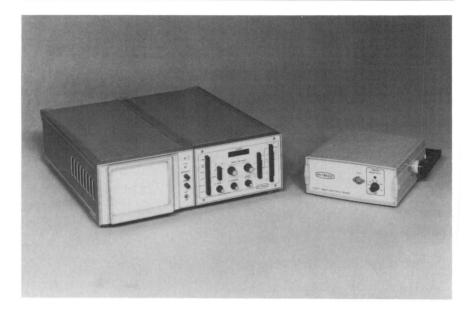

Fig. 9.5 The T276 (*source:* Biccotest Ltd, UK)

In order to provide pre-trigger information, the trace patterns for the last group of pulses are held in memory and constantly updated. The unit is triggered by a voltage sensitive device of variable sensitivity when the fault arcs over and a voltage depression occurs, such as that shown in Fig. 1.7a.

Even over this relatively short time span several pulses will have been injected during the period of voltage depression, while others will lie before and after this point. Thus several traces have been memorized, some before, some during and some after the fault arc. A selection feature then allows the operator to switch through and view each memorized trace and superimpose and compare a trace for the faulty state (during the depressed period) with a trace for the healthy state (on an undisturbed point on the sine wave).

The P240, though mostly used for normal PE/TDR tests on dead or live low voltage systems, has a remote control facility enabling it to be triggered from an external source. When used in this mode, the trigger is activated by either a low voltage detector, a current sensor or rate of change of voltage detector and the last trace is memorized. This is then displayed superimposed on the healthy state trace and a point of diversion again indicates the location of the fault.

It must be said here that fault location by this method is not just a case

of connecting the unit and inserting a new fuse to re-energize the circuit. Several factors mitigate against success.

The first of these is the sheer unpredictability of the low voltage fault! The fault may have become permanent, in which case the set will trigger immediately and produce traces of a faulty state only. It is much more likely, however, to be of a transitory nature, blowing fuses when *no* equipment is connected and not manifesting itself at all for long periods when equipment *is* connected! This, though, is the nature of fault location on loaded systems and it has to be accepted that a certain percentage of faults will always resist detection.

A further problem has to do with the means of re-energizing a faulty cable with a transient fault locator connected. If no special measures are taken to avoid it, the pulses from the set will not only travel up the cable but also into the low voltage bus bar system and thence up a multitude of parallel paths, i.e. other connected cables. Such a low impedance 'reservoir' for the pulses can make a nonsense of the traces obtained in that they will not be solely indicative of the cable under test.

It is common practice therefore to use a modified fault re-energizing device for energizing the cable. The modification involves introducing an inductance into the 'take-off' lead of the re-energizer to act as a blocker so that the preferred path for the pulses will be into the cable and not back into the bus bar, as shown in Fig. 9.6.

The inductance is simply several turns of the heavy duty flexible connection cable, tightly bound together and restrained to counter the large physical forces that would otherwise throw the coils apart when fault current flowed. Thus, yet again, the need for re-energizing devices is demonstrated.

SOUND APPROACH

Low voltage fault location, more than any other branch of the art, demands *fault location management*. A fault on a high voltage cable several kilometres

224 UNDERGROUND CABLE FAULT LOCATION

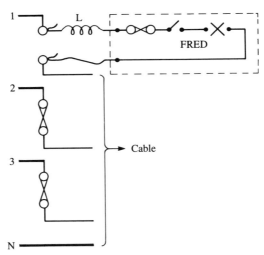

Fig. 9.6 Schematic showing inductance in use with fault re-energizing device

long is usually much easier to find – all the necessary equipment will be available, the cable is clean and disconnected, there is no serious limitation on applied voltage, no consumers are off supply, experienced staff are involved, etc.

The low voltage fault on a loaded system is a different matter altogether. Faults occur daily and the hard pressed staff who attend may or may not be well trained and well equipped and, in addition, they have to face very serious pressure to restore supplies in the shortest possible time. The nice guy standing by the side of the joint hole politely enquiring about the incident . . .

. . . becomes another creature altogether when he decides that the supply company has intentionally turned off his supply during his favourite television programme.

At the very top of the list of basic requirements are, of course, sound management policies which ensure that enough dedicated staff are quickly directed to the trouble spot, fully equipped with the latest instruments and supported by staff and resources for the necessary digging, jointing and provision of temporary supplies, etc. Although every authority and company strives to get close to this ideal, such perfection is rarely achieved and the best approach to low voltage fault location is to combine the application of whatever resources and equipment are available with a logical common-sense approach in order to achieve restoration of supplies in the shortest possible time and with minimum damage to the system.

On arrival at the substation fully armed with the relevant network

POWER CABLES 225

diagrams and plans, the engineer(s) should use filament test lamps or a voltmeter to diagnose the fault condition as already described. This will indicate what fuses are ruptured, if any, and whether there is a core-to-core condition or, in conjunction with any available intelligence regarding consumers 'off supply', if a probable open circuit situation exists. In any case, the next step is to record at least the faulty core trace in PE/TDR memory before the cable is re-energized.

An audio frequency signal should be applied to the cable and the route traced and marked (see Chapter 6). While walking the route, the operators should carefully collect and note *all available evidence*. This may comprise sites of recent excavations, the comments of consumers, evidence of unusual heating, e.g. melted snow, etc.

At the beginning of the subsection on 'field turbidity method for core-to-sheath faults' it is suggested that a suspect zone may be indicated while tracing the route. This will be so providing the audio frequency generator is connected between the faulty cores and earth, as is often the case. No such indication will be noticed, however, if the generator is coupled inductively via the transmitter coupling tongs or aerial.

It is clear that one person can carry out these checks and tests alone, but two people are much more effective; methods and evidence can be discussed, one person can check and mark plans while the other is tracing, two-way radio can be used to great effect when one engineer is perhaps re-energizing in the substation with the other stationed along the route, etc.

At this point in the proceedings pressure is already beginning to mount to 'get supplies back'. Some people have been off supply for some hours, since the fuse first ruptured, and some for around an hour, these being the consumers on the other phases disconnected so that testing could be carried out.

It is now time to re-energize. This will be done as much to obtain a pre-location as to restore supplies. The method of pre-location depends, of course, on what equipment is available – a PE set/TDR alone or a transient fault locator plus a fault re-energizing device. Taking the latter situation first, the transient locator should be set up to view the faulty core (as yet still dead). The 'healthy' phases should now be re-energized by inserting fuses. This will restore most supplies and establish the electrical conditions ready for the re-energizing of the faulty core.

This can now be done using the fault re-energizing device (FRED) after first making sure that the transient fault locator is armed. The FRED will trip out immediately, or after a short time, or not at all. If a trip occurs, the transient fault locator will trigger and a trace such as that in Fig. 9.7 will be obtained, thus allowing the fault distance to be measured.

It is more likely that the FRED will not trip immediately but will 'hang

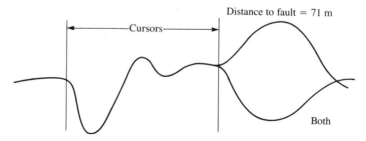

Fig. 9.7 Trace obtained when transient fault locator is used with a fault re-energizing device

in' for quite some time after being closed. If it does eventually trip, then the situation will be as just described and a pre-location can probably be made. If it does not trip, however, there is still a very good chance that the transient fault locator will trigger due to one of the many transitory fault arcs such as one of those shown in Fig. 1.7b. These occur often but are short lived and do not cause a fuse to blow.

If only a PE set/TDR is available, the faulty phase is best re-energized by inserting a fuse. When this fuse ruptures, the pulse echo set should be connected again to see if the fault arc has modified the trace.

One of four situations now exists:

1 There is a pre-location and some supplies are off.
2 There is no pre-location and some supplies are off.
3 There is a pre-location and all supplies are temporarily normal.
4 There is no pre-location and all supplies are temporarily normal.

In the first case pin-pointing can be attempted at the point(s) indicated by the pre-location or at some close-by suspect area. It should be mentioned here that the surface or ground at these points can be tested for temperature or the presence of the typical fault smell. Evidence of warm ground can be melted snow or dry patches on an otherwise wet pavement. Checks can also be made with the palm of the hand or by 'more technical means'.

Infra-red imaging equipment will easily show up any 'hot spots', but this type of equipment is, of course, too expensive for deployment on a day-to-day low voltage location. If available, a specially adapted thermal 'gun' is most effective. Such a device is described in Chapter 11, 'under-floor and under-road, pipe and soil heating systems'.

Every experienced operator engaged in fault location knows the characteristic smell produced by a faulty cable burning underground. This smell permeates the earth surrounding the fault and it can be sampled overground quite easily by driving a large screwdriver or spike about 300 mm (1 ft) long down into the ground at some convenient point such as a gap between paving slabs. This approach is certainly not very technical but who can argue with a method that has pin-pointed so many faults within minutes and without using instruments? Many an inexperienced engineer, standing in a substation beside a pile of spent fuses, has been surprised by the supervisor appearing and telling him exactly where the fault is!

The supervisor has walked the route, perhaps talked to a few people, perhaps seen signs of recent excavation, perhaps tested for smell, etc. This further emphasizes the validity of the list of manifestations given at the beginning of Chapter 4, 'pin-pointing'.

Normally speaking, though, the main pin-pointing method used in low

228 UNDERGROUND CABLE FAULT LOCATION

voltage fault location is to station people at suspect points, with or without acoustic detectors, and repeatedly re-energize the fault, preferably with a FRED (some versions of which will automatically reclose several times before locking-out). If, by luck or good management, one of these people happens to be standing over or near to the fault, the resulting sound/vibration is usually obvious as a great deal of energy is dissipated. The 'banger' in this case is not an instrument but the supply system itself.

If the fault is conclusively pin-pointed, well and good. If not, an excavation will be made at one of the pre-locations and the cable inspected. A PE/TDR pre-location is, of course, a distance derived from a pulse 'running time' in microseconds, and can indicate more than one possible location on a multi-branched system. Reference can be made here to Fig. 3.34 and the associated comments relating to 'running time', e.g. a fault at 4.5 μs could be on the main between x and z or up the tee branch x–y.

The fault will then be confirmed. If there is no sign of a fault, the cable may be cut and the PE set/TDR applied at this point, looking both ways into the cable. If the fault is close by a large trace feature will give a good location.

Once a cut has been made each half can be re-energized and supplies restored on the healthy half. The fault in the other half will then be found by technical means as described or by further cutting and testing.

In case 2 where no pre-location has been made, it will be necessary to take a chance on repeatedly re-energizing the cable and listening at various joint positions and suspect areas. Should this fail, it will unfortunately mean resorting to 'cut and test', owing to mounting pressure to restore supplies, not only from consumers but also from management.

Situation 3 is interesting. There is a fault which has ruptured a fuse. It has apparently 'gone away' but is, of course, still lurking and will strike again, probably at midnight on a holiday. A pre-location is available, but there is no easy way of confirming it. Depending on the time of day or night and the availability of resources, it may or may not be worth while carrying out excavations and inspecting the cable. At worst, records can be marked in preparation for the next time a fuse blows or the procedure for situation 4 can be carried out as described below.

In cases like situation 4 and situation 3 above, it is a shame to leave the scene without doing something constructive and simply await the next occurrence. The very sound basic tenet of always trying to capitalize on the behaviour of a fault exists here. In this particular situation the fault will arc over again and it is good practice to use this occurrence to obtain a pre-location or at least advance the fault location process somewhat.

Should a FRED and transient fault locator be available, they can be left in circuit and checked periodically. If the FRED trips there is a very good chance that a pre-location trace will have been captured. Even if it does not trip, occasional fault arcing may well trigger the transient fault locator and give a pre-location.

If this equipment is not available, there is one constructive action that will pay dividends; the cable can be sectionalized so that the next time a fuse ruptures a section of the previously connected network will have been declared healthy. On an interconnected network this can be effected by changing a split point or, if no link boxes or pillars are in circuit, the cable can be cut and temporary stop end joints made. Even on a radial feed, the cable can be cut and low value fuses temporarily inserted.

9.2 Faults on medium and high voltage systems

TYPES OF SYSTEM AND INFLUENCE ON CHOICE OF METHOD

It can be argued that 'a cable is a cable and a fault is a fault'; that is to say, a pre-location instrument will not know (or care) whether it is connected to a telecommunications cable or a high voltage cable. However, the make-up of a cable or cable system often dictates not only what type of fault occurs but also what method or approach needs to be employed to locate a fault. Comment now follows on the following types of cable system:

- MV and HV three-core oil-impregnated-paper cables with no tees
- MV and HV three-core plastic-insulated cables with no tees
- MV and HV single-core cables laid flat or in trefoil, with no tees
- MV and HV cables with tees
- Unearthed MV and HV systems
- EHV cables
- Oil and gas filled cables

MV and HV three-core oil-impregnated-paper cables with no tees

Faults on these cables are usually the most straightforward to find. Mainly, the cables differ from those of lower voltage only in the aspects of length and voltage withstand. They can be isolated at each end and the paper insulation will usually burn down if stressed.

Therefore all the pre-location methods already described, resistance bridge, PE/TDR, impulse current and arc reflection, can be used. The cables are rarely too long for PE/TDR. Pin-pointing by shock wave discharge is extremely effective because high voltages and energies can be utilized. Audio frequency methods are not normally used or needed.

However, there is one situation that can arise on such systems and causes considerable difficulty. This is the occurrence of a fault that trips the circuit but withstands full d.c. test voltage thereafter. This is often due to the fault having occurred in a joint filled with a mobile dielectric such as oil or compound. The fault arcs over, carbon is formed but, after the explosion, the dielectric moves in to reinsulate the fault path.

It is not uncommon for such a cable to be reclosed on to the system, whereafter it remains alive and apparently healthy for some time. However, eventually stresses cause the particles of carbon to realign and the fault arcs over again.

Frankly, if a cable withstands full pressure test voltage for the prescribed period, it is healthy at that point in time and only two courses of action are open: the Chief Engineer of the authority concerned can authorize a higher test voltage or a longer period or, as is normally the case, the cable is switched in again and left energized but off load until the fault begins to recur more frequently. There are also some possibilities of testing for the presence and sites of partial discharge. These are set out later in the section on 'current practice'.

When one of these very awkward faults is finally failing occasionally under a pressure test the best and almost the only approach is to rig for impulse current pre-location using the 'relaxation' method (see Fig. 3.57 and the associated text).

MV and HV three-core plastic-insulated cables with no tees

Most of the comments in the first category above apply to this group of cables except that, because burn-down is not usually possible, most pre-locations are carried out by transient methods.

MV and HV single-core cables laid flat or in trefoil, with no tees

This category represents the systems most in use today that utilize three single-core plastic-insulated cables laid side by side or bound together in trefoil. Each core has a screen so that core-to-core faults cannot normally occur. Fault location is as in the previous category above, although the location of faults on the individual sheaths/screens is often carried out if the system construction allows them to be disconnected from earth at each end. Sheath fault location overground is fully covered in chapter 4, 'pin-pointing', but there are methods of sheath fault *pre-location* which are discussed at the end of this chapter under the section on 'current practice'.

MV and HV cables with tees

Be they paper or plastic, three-core or single-core, high voltage circuits with tees make the fault location engineer's life pretty miserable! There are some circuits with just a few tees and some that have many tees. At the end of every tee is a load – a transformer. It can be connected directly or via an isolating switch or plug-in device. These transformers have to be disconnected from the cable system before any meaningful fault location can be attempted.

Although the use of tees is very cost effective, it immediately introduces three factors that make fault location extremely difficult:

1 All methods of pre-location are made more complicated.
2 Each extremity must be visited to disconnect the transformers and several must be visited to make connections during pre-location.

3 The very fact that loads are at the ends of tees and not fed from disconnectable sections on rings means that the fault has to be found and repaired or cut out before any supplies can be restored. This is in direct contrast to a fault location on an isolated section of HV cable in a ring system, where all supplies have been restored by switching.

Therefore, not only are fault location tests trickier and more time consuming, but there is also extreme pressure to reach a successful conclusion and restore supplies. Fault pre-location by resistance bridge on teed systems is fully explained in Chapter 3 in the text associated with Fig. 3.17 and involves applying a strap at two or more extremities in order to resolve the situation.

An analogous approach when using pulse echo pre-location also necessitates making checks at two or more ends because any one 'time' or 'distance' reading can indicate fault positions at two or more sites, as already explained under low voltage fault location with reference to Fig. 3.34.

The most onerous situations are found on some 20 kv circuits, as mentioned above, where there can be up to 100 tees along a cable many kilometres long. It is clear that disconnecting transformers and carrying out fault locations would be impossible if supplies depended on one cable only. Therefore it is normal for two cables to be installed, an 'a' and a 'b' circuit so that, if either fails, all the transformers can be switched over to the other cable to restore supplies while the engineers find the fault on the first cable.

Necessity certainly is the mother of invention. The engineers required to troubleshoot on such systems have become expert at interpreting complex pulse echo traces and have also adopted certain unorthodox but most effective ploys in finding faults on them. In the past one such ploy was to use powerful high voltage burning equipment to burn faults down to zero ohms or at least a very low value. Because of the complexity of the system, the trace will still be difficult to interpret, even using 'before and after' and core comparison methods. The method then involves deploying a person to various substations along the route to apply a repeating short circuit. If this 'flashing earth' is applied before the fault, it is visible on the screen. Wherever it is applied at some point after the fault it ceases to be visible.

Nowadays, tests are usually carried out with impulse current equipment using the 'loop on, loop off' technique, the straps being connected on and off at several extremities, not at the test end, (see Figs. 3.63, 3.64 and 3.65 and the associated text). Nonetheless, the 'flashing earth' approach is a useful idea for teams that do not possess refined impulse current equipment.

Unearthed MV and HV systems

The main problem found on unearthed systems is that most faults encountered are high resistance ones. This comes about because an earth fault occurring on one phase simply locates that phase at earth potential. No fault current flows but the high impedance monitoring equipment located at the source substation indicates the presence of a fault and on what phase it lies. The circuit is then disconnected and the fault found. No major arcing has taken place so the fault is invariably a high resistance one.

As a matter of interest, these unearthed systems need to be maintained in good condition, particularly where conductors and insulators can be exposed to dirty or damp environments. The reason for this is that an earth fault on one phase raises the phase to earth stress on the other two phases to the phase-to-phase voltage value, thus overstressing all insulation on these phases.

If the circuit is not shut down, a so-called 'cross country' fault can occur if another phase breaks down to earth. A phase-to-phase fault then exists and massive fault currents flow.

EHV cables

This heading is intended to point up the category of flashing faults with a breakdown threshold higher than the voltages available from any available surge generator or burn-down set. This usually means above about 32 kV. Straight away, methods such as resistance bridge, PE/TDR, arc reflection and conventional impulse current are ruled out.

Because the fault needs to be subjected to a higher voltage to manifest itself, the method employed *must* utilize a d.c. pressure test set with a high enough voltage to break the fault down. As no arc reflection equipment can as yet cope with voltages of, say, 70 or 80 kV, the method that has to be used is impulse current. There are two possibilities, both of which use the 'far end loop on, loop off' method, as it is impossible to utilize double test leads and near end loops when operating at very high voltages.

A purpose built test van may be available with a high voltage d.c. source of around 80 kV associated with a fully rated capacitor bank and spark gap. This then constitutes a very high voltage surge generator and a 'far end loop on, loop off' test can be carried out with the linear coupler in the earth return. This will give results comparable with those shown in Figs 3.63, 3.64 and 3.65. The only difference is the higher voltage and equipment ratings.

The other possibility is to use a healthy cable core as 'surge capacitance'. The circuit configuration for this is shown in Fig. 9.8 where a high voltage

234 UNDERGROUND CABLE FAULT LOCATION

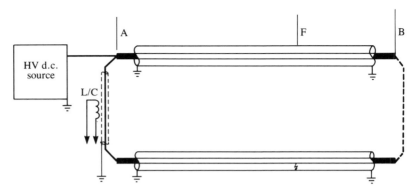

Fig. 9.8 Connections for impulse current test using high voltage

d.c. test source is connected to faulty and healthy cores which are commoned by a special loop along which the linear coupler is clamped.

This loop is simply made up of a length of fully rated screened cable with the screen at each end cleanly cut back and sleeved clear of the core ends. A single earth connection is run from this screen. This ensures that there is an earth between the linear coupler low voltage 'winding' and leads and any high voltage, thus making the rig quite safe to use.

The d.c. test set is then switched on and the voltage raised until the fault breaks down. When it does so the impulse current set is triggered and a 'relaxation' trace memorized (see the text associated with Figs 3.51e and 3.57). The circuit is then switched off and earthed and a loop applied between the faulty and healthy cores at end B. High voltage d.c. is again applied until the fault breaks down.

The transient starting from the fault site and travelling towards the source end produces an identical trace to the one just memorized. However, this time, the *other* transient, which was previously trapped between the fault and the remote end B, is free to travel through the far end loop and arrive at the linear coupler to produce the feature marked 'breakdown 2' in Fig. 9.9, which depicts the two traces superimposed.

Note on Fig. 9.9 that if a 'strange trace' is obtained, it is probable that the linear coupler has been clamped on the wrong way round. The resulting trace is simply the above trace inverted. The point of diversion will still be obvious.

Thus there are two breakdown features marked 'breakdown 1' and 'breakdown 2', separated in time by the time lapse T_f. This time lapse is quite simply the extra time taken by breakdown 2 to 'go the long way round' and is equivalent to twice the distance FB. As the propagation velocity factor, $V/2$, is always used, this means the distance to the fault

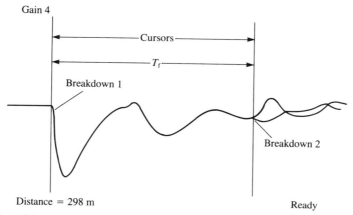

Fig. 9.9 High voltage impulse current test – 'loop on' and 'loop off' traces superimposed

from the far end. Even though quite complex traces are encountered, this point of difference is all that matters and, if this is visible, interpretation and measurement are straightforward.

Oil and gas filled cables

Core faults on very high voltage cables insulated with the mobile fluids, oil and gas, are very rare. Such cables are divided into oil or gas 'sections' by oil or gas 'stop joints' and, of course, the terminations where applicable. The insulation fluid pressure in these sections is permanently monitored.

If a hole appears in the metal sheath the oil or gas escapes and the resulting loss in pressure on the section concerned is automatically signalled to a main control centre. If the loss is minor, fluid can be pumped into the section to maintain pressure while steps are taken to locate the site of the leak. There is very sophisticated mobile equipment available that can monitor pressures and loss rates at both ends of a section and compute a pressure gradient for the section that indicates the approximate location of the leak. This is very expensive and not universally available.

The more normal approach on oil filled cables is to dig down, expose the cable and apply a special jacket to it to freeze the oil, thus creating a block to the flow. Doing this at several points along the route enables the cable to be sectionalized until the suspect section is narrowed down to a short length (the equivalent of 'cut and test' without cutting!). This is also quite expensive as it requires special 'oil vans' with the necessary equipment built in. Such vans are usually owned by large cable and contracting companies and some of the transmission and distribution companies.

236 UNDERGROUND CABLE FAULT LOCATION

By far the best approach is to make sure, by preventive maintenance, that holes never normally develop in the metal sheath. This sheath is always protected by a plastic oversheath to prevent corrosion. The metal sheath is earthed by removable links so that it can be disconnected during maintenance and electrically pressure tested against earth to check if there is any damage to the plastic oversheath. If this is done on a regular basis, holes in the plastic covering can be found and repaired before any significant damage has been done to the metal sheath. Therefore, in theory at least, there should rarely be a fluid leak and there will never be an electrical fault (except from the failure of joints or major mechanical damage to the whole cable).

Sheath fault location is fully described in Chapter 4, 'pin-pointing', but there are also ways of pre-locating them, not as yet discussed. The traditional approach using a resistance bridge can be effective, the most favoured method being the 'Hilborn loop' already described. For this, the two auxiliary cores utilized can be two healthy cable cores or, as cross-section is not critical, spare pilot or telephone cable cores terminated in the substations at the two cable extremities. It should be remembered that such sheath fault tests are usually being carried out on high voltage cables where terminations are outdoors on high level structures in compounds and switchyards, so that matters of access and convenience are always factors to be considered.

A major disadvantage with these tests can, however, be induced voltages and interference due to the presence and proximity of high voltage, heavy current circuits, earth currents, etc. A more modern approach is now provided by several manufacturers which utilizes the small high voltage burner/d.c. source which, in any case, will be used for the 'pool of potential' pin-pointing to be carried out after pre-location. The test is analogous to the Hilborn loop resistance bridge test because it also requires two spare cores as auxiliary connections. Instead of comparing resistances, the method is to compare voltage drops in two sections of the sheath under identical current flow conditions. Figure 9.10 shows the test circuit and connections. In this illustration, three single-core cables are shown. The faulty sheath is on cable 1 and the cores of cables 1 and 2 are chosen as the two auxiliary connections. If the faulty sheath were on a three-core cable, any two cable cores could be used in the same way. Basically, as in the Hilborn loop test, any convenient cores can be used.

The principle of the test is as follows. With the changeover switch in position 1 a constant direct current flows along the section of the faulty sheath marked L_1, through the fault and back to the d.c. source via earth. The voltmeter V registers V_1. It should be noted that the faulty sheath section L_2 is not carrying current from the d.c. source but is simply acting as a voltmeter test lead. V_1 is noted.

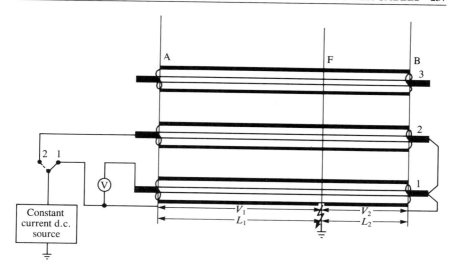

Fig. 9.10 Connections for sheath fault pre-location test

The circuit is now fed by the d.c. constant current source with the changeover switch in position 2. This time, the constant direct current flows along core 2, through the faulty sheath section L_2 and to earth again via the fault. In this case the faulty sheath section L_1 carries no test current and becomes part of the voltmeter test lead. The voltmeter now reads V_2.

The reversal of polarity between the two tests is of no consequence mathematically; nor is it inconvenient if the voltmeter is a digital one, as is usually the case. As the current is the same in tests 1 and 2, the voltage drops V_1 and V_2 are comparable and proportional to sheath resistance and therefore length. The fault distance from end A is

$$\frac{V_1}{V_1 + V_2} \times L$$

This approach, utilizing a source with a power of about 0.5 kW and capable of applying several kilovolts, does away with problems of interference and produces significant and measurable voltages in millivolts.

The best accuracy attainable is approximately 1 or 2 per cent but this is quite satisfactory as these earth contact faults are easily pin-pointed by the pool of potential method already described.

HISTORICAL APPROACH AND CURRENT PRACTICE

Traditionally, engineers engaged in high voltage fault location have had at their disposal a full set of equipment, more often than not installed in a tailor-made test van or trailer. They were, and still are, well trained and expert in the use of this equipment and able successfully to carry out fault locations using PE/TDR and, in recent years, impulse current methods of pre-location, together with all the normal methods of pin-pointing described herein.

The impulse current method has its devotees; people have been weaned off PE/TDR as a basic approach and use impulse current as the first line of attack because it can cope with all types of fault, including the flashing fault, without the need for treatment. On the other hand, a large body of engineers has been using an approach based on PE/TDR pre-location throughout the same period. With regard to pin-pointing, this has basically changed little over the years although refinements in equipment design have been considerable.

The main impetus for change has been in the realm of pre-location and has been provided by the advent of transient fault location in the form of impulse current. This has produced invaluable knowledge regarding the very anatomy of faults and related transient phenomena. Also, the truly wonderful capabilities of modern electronics in facilitating the capture, retention, display and analysis of transient phenomena which has come together with the development of the impulse current equipment has created the platform for current practice.

The combination of these two factors has made possible the development of the modern versions of the arc reflection method which has taken over as the most popular and effective pre-location approach. Moreover, the 'interface/coupling unit' at the front end of most modern equipment enables the fault location engineer quickly to carry out any sort of arc reflection or impulse current pre-location using the built-in HV sources in absolute safety and with a minimum of setting up, control and interpretation.

As stated earlier, arc reflection is a very persuasive technique which inspires total confidence in the operator because a recognizable change is able to be created in the trace at the point of fault and, into the bargain, no interpretation is required.

An Appendix follows that contains actual fault location examples submitted by several manufacturers. Each example is annotated and is self-explanatory.

Appendix 9A Examples of actual fault locations

AVO BIDDLE INSTRUMENTS, USA
Fault at Honey Creek Mall, Terre Haute, Indiana

Method: Arc reflection
Power utility: Public Service of Indiana
Location: Terre Haute, Indiana
Site: Honey Creek Mall (see Fig. 9A.1)
Cable type: 750 MCM (3φ system)
Cable length: 1360 feet
Conditions: Complete cable buried beneath 60 in of sand fill covered by 4 in of blacktop
Problem: The mall was plagued by an irregular power source.
Fault was high resistance and could not be seen by a TDR alone.
PSI crew had spent three unsuccessful days circling the mall parking lot trying to locate the fault with traditional methods (impulse generator only).
PSI was prepared to tear up the entire parking lot to replace the cable (estimated cost—$100,000).
Solution: Ron Troutman (AVO International Regional Sales Manager) brought the PFL-3000 Arc Reflection System to the Mall and located the fault (at 520 feet) in 20 minutes.

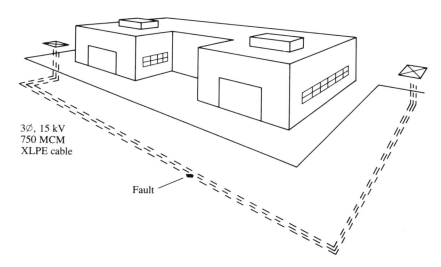

Fig. 9A.1 The site at Honey Creek Mall

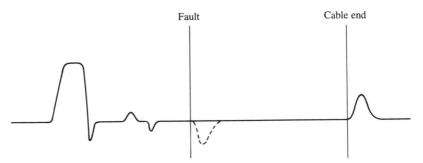

Fig. 9A.2 Arc reflection trace

Methodology: Used a Megger™ to determine that Aφ was good and B/Cφ's were faulted.

Used the TDR portion of the PFL-3000 Arc Reflection System to determine the actual velocity of propagation for the cable under test. (Note: velocity of propagation for a specific buried cable tends to change over time due to environmental factors.)

- Normal velocity of propagation for 750 MCM is 0.56.
- R. Troutman hooked system to the good phase and set the distance readout on the TDR to 1360 feet.
- He adjusted the velocity of propagation knob until the TDR cursor stopped at the beginning of the upward reflection that represented the end of the cable.
- Velocity of propagation for the cable under test was actually 0.48.

Using the true velocity of propagation of the cable under test, he then switched on the impulse generator and adjusted the cursor to the beginning of the momentary downward reflection (see Fig. 9A.2).

The distance to the high resistance fault was approximately 520 feet. Because the sand muffled the 'thump', the acoustic detector showed only a slight deflection at the exact location of the fault.

Additional data: Having an accurate localization (with the high voltage radar) was the only way to then pin-point the fault.

On the TDR's 2000 foot range, using 0.56 rather than 0.48 for the velocity of propagation would have resulted in an error of more than 100 feet.

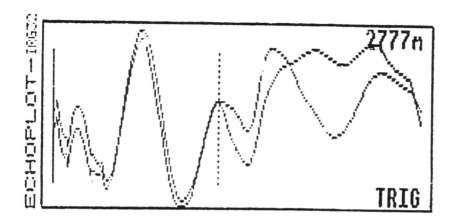

Fig. 9A.3 Secondary impulse trace

BAUR PRÜF UND MESSTECHNIK GmbH & CO, KG, AUSTRIA
BAUR TEST EQUIPMENT LIMITED, SUTTON, UK
Fault in Salisbury, UK
Method: Secondary impulse (equiv. Arc Reflection)
Cable: 0.15 in^2 Cu paper ins. 33 kV
Length: 3290 m
Route: Substation to pole box
Failure: Broke down in service
Equipment: Syscompact incorporating PE set/TDR IRG 32
Pre-location: Using surge generator and with delay setting of 2.5 ms, IRG 32 triggered to give display shown in Fig. 9A.3. Cursor set to separation point of initial trace and trace showing arc—reading to cursor 2777 m.
Pin-pointing: Actual fault position pin-pointed using surge generator and acoustic detector within 10 m of location.

BICCOTEST LIMITED, UK
Fault on a cable with multiple tees in London
Method: Impulse current, using 'loop on', 'loop off' at remote ends. All procedures, diagrams and calculations are shown in Fig. 9A.4.

SEBA DYNATRONIC GmbH, GERMANY/AVO INTERNATIONAL, UK
Fault in Troon, Ayrshire, Scotland. See Fig. 9A.5.
Method: Arc reflection
Cable: 0.06 in^2 Cu paper ins. 11 kV

242 UNDERGROUND CABLE FAULT LOCATION

Type of cable
6.6 kV 0.25(Cu) 3-core PLSWA

Circuit length
4589 metres multiteed

Calculated propagation velocity $\left(\frac{V_P}{2}\right)$
79.44 m/μs

Fault condition
Red phase intermittent fault to earth breaking down at 6 kV. Yellow and blue phases Ph/E and Ph/Ph 100 megohms plus

Test mode
D.c. loop comparison method

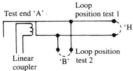

Test arrangement with equipment at 'A'

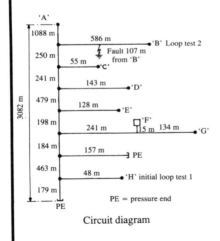

Circuit diagram

Test 1 (from 'A' to 'H')
23.45 μs from breakdown to separation of traces

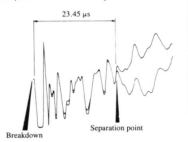

Trace obtained

Calculations
Distance from 'H'
 = 23.45 μs × 79.44 m/μs $\left(\frac{V_P}{2}\right)$
 = 1862 m

1862 m from 'H' indicates within 1 metre the tee joint of the 'B' leg.

A further test is required to establish whether the fault is at the tee joint or on the 'B' leg.

Test 2 (from 'A' to 'B')
1.35 μs from breakdown to separation of traces

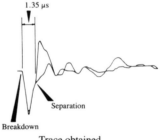

Trace obtained

Calculations
Distance from 'B'
 = 1.35 μs × 79.44 m/μs $\left(\frac{V_P}{2}\right)$
 = 107 m

Result
Calculated distance to fault from 'B' = 107 m
Actual distance to fault from 'B'   = 106 m

Fig. 9A.4 Network connection and traces for impulse current fault location on a multi-teed system

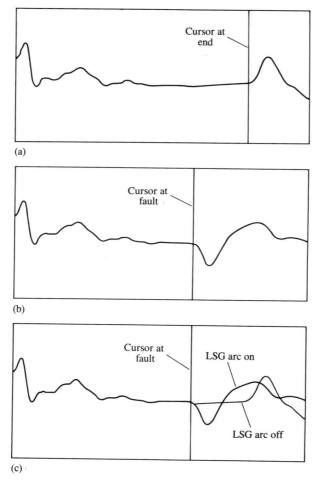

Fig. 9A.5 Arc reflection trace (a) LSG arc off, (b) LSG arc on, (c) (a) and (b) traces superimposed

Length:	640 m
Route:	Oil switch fuse to remote transformer
Failure:	Switch tripped due to 40 A fuse blowing in blue phase
Equipment:	Test van incorporating arc stabilization unit LSG and PE set/TDR/monitor CAF eta
Pre-location:	As in report below, showing traces for the cable condition before and during the arc together with the two together
Pin-pointing:	Actual fault position confirmed using surge generator and acoustic detector within 10 m of location.

244 UNDERGROUND CABLE FAULT LOCATION

11 kV fault location report

Report	: 4a	District: Ayrshire Date: 5/10/92		
Primary source: Troon		Section: Golf Crescent to Fullarton Drive		
Cable length	: 640 m	Phase : b Propagation velocity: 83 m/μs		
Pressure test	Breakdown kV	Breakdown obvious?		Pre-location length
Voltage Coupler?				m
Current Coupler?				
Arc stabilization	Surge kV	LSG operated	Split OK?	Pre-location length
	4	Yes	Yes	492 m
Impulse current	Surge kV	Loop used?	Split OK?	Pre-location length
	8	No	Yes	486 m
Burner	Used?:	No	Pulse echo location: m	
Remarks: lsg on mains		Engineer: I. Watt		

10

Telecommunications systems, information and control systems

10.1 Telecommunications systems

FAULT LOCATION PHILOSOPHY
In thinking about telecommunications fault location it is important to consider the very basic and distinct differences between telecommunications systems and all other systems except information systems. Although some of these differences may seem obvious when pointed out, they entirely dictate not only the choice of method and equipment but also the organizational approach.

For instance, a power cable transports a commodity, electricity, to consumers who share and tap into the same cable cores to draw it off. The telecommunications cable, however, carries a signal from one person or machine to another. This signal, moreover, is destined for the recipient only and travels along a dedicated pair of conductors which physically end up at the recipient's premises.

This two-pole connection has to be maintained electrically between the sender of the signal and the recipient over long distances via a multitude of joints and connectors of diverse types and ages situated in diverse environments. Herein lies the most significant difference between electrical distribution cable fault location and telecommunications fault location. The former is a matter of where the commodity is blocked or is leaking out of a short, massive cable with relatively few well-protected joints and connections. The latter requires the location of a point on a signal route at which a signal is lost, diverted, blocked or degraded. Furthermore, this route is made up of very small cross-section wire connected through many joints, cabinets and connector blocks (excluding of course all fibre optic and radio links). More than 90 per cent of faults occur at joints and connections.

Re-routing a signal path is normal in telecommunications work, whereas the provision of a temporary or alternative feed for a single power customer is not routine. While an electrical system fault is tackled by a power engineer who sits at the end of a dead and isolated cable equipped with instruments for diagnosing, pre-locating and pin-pointing the fault, a reported telecommunications fault condition is first classified and diagnosed by exchange/central office staff. After this a troubleshooting technician or team takes over to sectionalize the network and 'home in' on the faulty equipment or section of cable. For much of this process straightforward diagnostic instruments such as volt/ohmmeters are used.

Pin-pointing on power cables is carried out routinely whereas it is difficult and often impossible to carry out pin-pointing on telecommunications cables. All fault location operations on a power system tend to be undertaken by one well-qualified and trained engineer or technician while, in the telecommunications field, different tasks are carried out by several operators with different qualifications, training, skills and specialities.

A tremendous difference in techniques is dictated by *resistance*, both *fault* resistance and *conductor* resistance. An insulation resistance of 50 megohms on an HV power cable is a fault, but a telecommunications cable carrying speech frequencies will run quite happily at such an IR figure.

The difference between the conductor resistances of power and telecommunications cables is also extremely marked. For instance, the resistance of 1 km of 24 AWG (0.51 mm) copper wire is 86.8 ohms, while the resistance of 1 km of a 300 mm² copper power cable core is 0.06 ohms! This affects considerably any technique involving resistance measurement such as a resistance bridge method.

Resistance and volt drop measurements on telecommunications cables can be made with extreme accuracy and connection resistances are of little importance. The opposite is true of power cable fault location by such methods, where extremely careful attention must be paid to test connections and the test itself involves very small volt drops.

An obvious difference is *voltage*. On every electrical distribution cable core there stands a voltage permanently stressing the dielectric between it and other cores and ground. This stress not only reveals the fault early but often dictates the fault type.

Voltages on telecommunications cables are low. Faults fester for a long time and are only revealed by a break in transmission or a loss of quality. The highest voltage a fault is subjected to is the voltage subsequently applied by the insulation resistance tester (a test that is not always wholly indicative of the effect the fault was having on the system!).

Problems also exist with the application of PE/TDR techniques. Although even a long coaxial cable presents a wonderful trace, small diameter twisted pairs are very 'lossy' and the transmitted pulse is grossly attenuated. Despite the ranges available on modern sets, it is very difficult to detect much significant trace detail beyond one or two kilometres on the very small pairs.

Many fault location methods rely on the use of high voltage and power. With a few exceptions these tools are denied the telecommunications engineer because of the size and voltage withstand of the cables, the existence of protectors and the basic fact of life that all pairs of a telecommunications cable can rarely be cleared of all traffic and isolated.

The mental approach of the telecommunications engineer is different from that of the power engineer. The telecommunications engineer or technician has a very refined 'feel' for the type and possible location of a fault derived from voltage, resistance and 'kick' tests. 'Kick' testing gives an approximate measurement to an open circuit fault if the IR value of a conductor is not too low. Every conductor has capacitance both to screen and to adjacent conductors. Therefore the capacity up to a break is proportional to fault distance. The capacity bridge exploits this.

When an ohmmeter is connected to a conductor, that conductor is charged up to the ohmmeter battery voltage and the meter needle 'kicks' and falls back. A reversing switch may be used to reverse polarity and cause another 'kick', which should be of equal strength. The peak value reached is a rough indication of the distance to the break. In the absence of an open locator or PE set/TDR this practice is a useful test when used by the experienced operator.

The engineer also thinks in terms of ohms per foot, yard or metre of cable conductor. This is a very logical and uncomplicated approach, particularly when dealing with hybrid systems with copper and aluminium conductors and cores of different diameters.

Finally, allied to the problems encountered in pin-pointing is the fact that, in many countries, underground telecommunications cables are laid in ducts. This means that, on the one hand, there are many access points at which tests can be made, while, on the other, it is unnecessary to pin-point exactly as the faulty length between two manholes will be replaced.

TYPES OF NETWORK AND HISTORY

A telecommunications network can be broken down into three main sections:

- Main trunk
- Local junction
- Local distribution

248 UNDERGROUND CABLE FAULT LOCATION

The trunk network used to be made up of long composite trunk cables containing coaxial pairs and balanced pairs. Transmission was by carrier frequency, the balanced pairs running at 6/54 or 60/108 kHz, the frequencies being limited by cross-talk, and the coaxial pairs operating at between 0.6 and 60 MHz depending on the number of speech channels. Amplifiers were required every 1.5 to 15 km. Nowadays, the trunk network is fully automated and is made up of a mixture of coaxial and optical fibre cable. The latter now carries over 80 per cent of all trunk traffic.

The local junction network is basically the cable network connecting exchanges/central offices. In an urban society these cables tend to be between a few kilometres and, exceptionally, about fifty kilometres long, and used to be made up of loaded quads run at speech frequencies. Loading consists of 88 mH series coils spaced at 1.83 km intervals to provide adequate response at speech frequencies. Though some loaded circuits remain, the vast majority of them now transmit by pulse code modulation (PCM) and there is an increasing number of optical fibre junction cables.

The local distribution network is made up of the cables from the exchange/central office to the customer and it is on this network that almost all faults occur. Figure 10.1 is a diagram of a local distribution network.

The main cables leaving the main distribution frame (MDF) can be very large, carrying up to 4800 pairs in some cases. They were paper core unit twin and are now polyethylene unit twin, i.e. they are made up of groups (units) of twisted pairs and many are pressurized with dry air (less than 1 per cent relative humidity) at 9 lb/in² (620 mbar) and have an internal aluminium moisture barrier. (Many main cables are now fully filled as the use of pressurized cables declines.)

Pressurized cables make for relative freedom from cable faults. Even if the cable is punctured, egress of air prevents ingress of moisture. At a

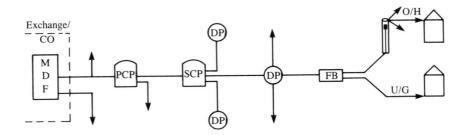

Fig. 10.1 Local distribution network

primary flexibility or cross connection point (PCP), a group of pairs is jointed out into the cabinet where the air pressure zone is terminated. The other pairs continue on to other cross connection points. Further distribution to secondary cross connection points (SCPs) is taken over by polyethylene multipair cables, most of these being 'fully filled', i.e. filled with petroleum jelly during manufacture to prevent ingress of moisture. They may or may not be screened and can be for underground use in ducts or manufactured as self-supporting aerial cables. A direct buried cable will be fully filled and armoured.

These cables terminate at distribution points (DPs) which can be in wall-mounted boxes and thence to underground footway jointing boxes (FBs) or pole-mounted connector boxes, whence the final connection to the customer is made either by underground cable or drop wire.

Along the route from the exchange/central office to the customer core diameters rise from a minimum size of 0.32 mm in the very large main cables to 0.9 mm further out on the system to compensate for transmission loss. The majority of routes are less than 3 km in length and the final leg of a route is a small cable of between 100 pairs and 2 pairs.

As 90 per cent of faults occur on this route, i.e. in the local loop, the most fault prone section lies between the cross connection point and the customer interface, there being some sort of connection/joint every few hundred metres and many vulnerable small cables at the ends of the routes. In fact, 90 per cent of all faults on cables occur on those with less than 100 pairs.

Much of the system in many countries is over 50 years old. Although copper predominates, there is still some aluminium and aluminium alloy cable installed in the late sixties and early seventies which is prone to problems arising from corrosion and breakage. In very dense built-up urban areas optical fibre can be taken right up to the customer interface and trials are being carried out in taking fibre all the way to the customer's premises on normal housing estates. However, copper still predominates and the requirement for conventional fault location on metallic cables will continue for a long time.

FAULT MANAGEMENT – INITIAL DIAGNOSIS IN THE LOCAL LOOP

For speech transmission the maximum attenuation in the local loop should be 10 dB at 1600 Hz. The loop resistance requirement for signalling from the exchange/central office to the customer is a maximum of 1000 ohms and the insulation resistance should be better than 2 megohms.

A fault is any deterioration in loss, continuity or IR which causes the circuit to fail or be unusable. This could be caused by ingress of moisture, a noisy connection, a break, a short, splits causing cross-talk, corrosion or

250 UNDERGROUND CABLE FAULT LOCATION

foreign battery (see the section on 'types of fault and influence on choice of method' on page 251). The fault may be reported by the customer, after which a report is raised and a test made from the exchange/central office, or a fault may be detected during routine remote testing overnight.

In either case the first action is remote testing of the line up to the customer interface whereby the type of fault can be diagnosed and, sometimes, the fault zone ascertained. A more thorough check can be made when automatic loop-back equipment has been installed at the (typically industrial) customer's premises. This enables a loop to be made or a tone injected at the terminal equipment without the need for a visit. The general aim of remote testing is to glean enough information to enable a faultman, lineman or faultman/jointing team to be despatched as near to the trouble spot as possible. This is the point at which 'fault location' as discussed herein really begins. That is to say, a trained and equipped troubleshooter is sent to check the circuit electrically and find the fault.

STAFF INVOLVED – TRAINING

Although practice varies in different countries, it is generally true to say that testing and fault location are undertaken by engineers and technicians of several levels and abilities. As stated earlier, they have very different qualifications and specializations.

At the top level are qualified engineers who are at home with loss and level testing, pulse echo (PE) time domain reflectometry (TDR), optical attenuation testing and working with optical time domain reflectometers (OTDRs). At the 'coal face' are technicians, classed as faultmen, linemen and jointers (hereafter faultmen), who tackle the vast majority of faults on the system (see previous comments on the percentage incidence of faults). In between these levels comes the engineer who is familiar with resistance and capacitance bridge tests as well as PE/TDR testing. This engineer is called in to sort out the trickier fault situations.

The faultman is familiar with all aspects of the local loop, i.e. types of cable, connectors, joints, cabinets and terminal equipment and has undergone thorough training in general testing, tone tracing and core/pair identification as well as fault diagnosis and location with a tone set or volt/ohmmeter/kickmeter. He or she may or may not have been trained in the use of a resistance/capacitance bridge or open, split and short locator, and was in the past unlikely to have used a PE set/TDR.

Over the last 20 years, a great deal of attention has been paid to fault management, including reporting, recording, remote testing/diagnosis and despatch time saving. However, reliance for the final location has rested on the well-trained and, until recently, very basically equipped faultman with a good 'feel' for the system.

TELECOMMUNICATIONS, INFORMATION AND CONTROL SYSTEMS

Advances are now being made as a new generation of 'clever' hand-held instruments arrives on the scene and people are being trained to use them. More importantly, the faultman is now more interested in and less suspicious of this new technology and is accepting the new instruments and using them to great effect in the companies and authorities in which they have been introduced.

TYPES OF FAULT AND INFLUENCE ON CHOICE OF METHOD

From Chapter 1 the types of fault found mainly on PTT/telecommunications systems are:

- Contact or short
- Break or open
- Ingress of moisture
- Crimping
- Cross-talk
- Ground contact

If a short is zero ohms, it can often be located by checking the resistance to fault with the ohmmeter. It will certainly succumb to a resistance bridge or tester such as those in Figs 3.8, 3.10 or 3.11. If available, a PE set/TDR will also see it clearly. In other words, it is not too difficult to find and there is a wide choice of method.

When the contact fault has some substantial resistance, the PE/TDR approach is ruled out. Also, a pure loop resistance measurement with an ohmmeter is meaningless in terms of fault distance because the core resistance is totally swamped by the fault resistance. Within limits the more refined resistance bridges will still be effective, although balance sensitivity, and therefore accuracy, is lost if the fault resistance is several megohms. Some of the hand-held devices mentioned above are also limited.

At the higher fault resistance values of, say, 2 to 50 megohms an inverted Murray loop resistance bridge or high impedance fault locator such as that in Fig. 3.21 is accurate and effective. Such instruments, though, are not in everyday use in PTT/telecommunications work. This is mainly because, historically, an IR value of more than 2 megohms has not been considered a serious fault. Also, the possession of different resistance bridges and a knowledge of different bridge configurations is necessary. This was simply not the case. However, the advent of small, reasonably priced hand-held instruments, such as the one referred to above, creates distinct possibilities. In addition, the quality and IR level of a pair is much more critical for digital traffic than for transmission at speech frequencies.

There is another area in which the high resistance fault creates problems, both procedural as well as technical. That is the use of a 'burn-down set' or

252 UNDERGROUND CABLE FAULT LOCATION

'fault enhancer' which produces current at the point of fault to lower its resistance. Though this practice is common on power systems, there are many factors that prevent or curtail its use on PTT/telecommunications systems, for instance:

- Traffic on other pairs within the cable
- Concern about the application of too high a voltage
- The presence of protectors which break down at a certain voltage

There is, however, no legitimate objection to the application of such an instrument at, say, a cabinet, pedestal or DP on a small pair plastic cable which is disconnected at the other end, if the applied voltage is kept lower than the test voltage of the cable. A successful burn-down will enable the fault to be found by PE set/TDR.

The awkward question that can be posed, though, is 'Will the faultman have a burn-down set or enhancer or, indeed, a PE/TDR set? Many of these problems and reservations are largely overcome by a new method whereby a high voltage is applied to the fault for a very short time. This allows a detectable audio frequency current to flow. This technique is set out in the following section on 'Methods and instruments – current practice'.

Ingress of moisture occurs frequently on telecommunications cables and it is not uncommon for ducts and manholes to be flooded. As indicated in the text relating to Fig. 1.4, the point of damage in the cable is where the water enters and it can be wet some considerable distance away from this, depending on the lie of the cable. Location of a wet stretch in a cable is not easy. On a PE set/TDR it can produce a trace like the one in Fig. 10.2. The capacitance of the cable dielectric is increased by the presence of moisture and the consequent change of characteristic impedance at the start and finish of the wet section gives the features shown.

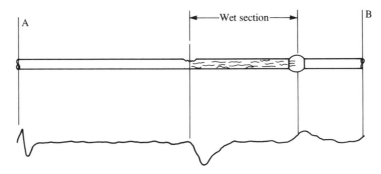

Fig. 10.2 The trace produced on a PE set/TDR when locating a wet stretch in a cable

The comments already made under 'traces found in practice' in Chapter 3 are worth repeating. In short, the transmitted pulse is progressively attenuated the further it travels. The change of characteristic impedance produced by a wet section is not very dramatic so the trace in Fig. 10.2 may be quite indicative if the affected area is close to the test end but invisible if far away.

'Close' and 'far away' cannot be easily defined, as the effect of attenuation depends on what mismatches, such as joints or line taps, occur before the faulty zone. Also, as already stated, 'distance' really depends on the losses in the pair being viewed. One kilometre is a long way on a 0.32 mm diameter twisted pair but is a very short distance on a coaxial pair.

The PE/TDR approach is best when all the cable cores are affected. If only some are affected, then a resistance bridge, normal or inverted, should be used. Typical IR values for damp cable are around a few kilohms. It is unusual to find one pair or only a few pairs reading low while the remaining pairs are showing very good IR values. The much more likely scenario is that readings will vary from a few kilohms to several tens of kilohms. In this context, the terms 'faulty core' and 'healthy core' need examining.

If the ratio of healthy core to faulty core IR values is greater than 1000:1, a bridge type test will be quite accurate. It is still worth while carrying out a test even if this ratio is around 100:1, although a significant error must be expected. It is impossible to be absolute on this matter because both the core resistance and, particularly, the actual fault position are among the factors influencing the error. Any 'bridge type' test, of course, always gives a balance or reading but, if the ratio is too low, the result may be inaccurate or almost meaningless. The graph shown in Fig. 10.3 gives a rough idea of variation in accuracy with change in healthy/faulty core fault resistance ratio. Both curves are for faults at approximately 50 per cent route length. It is interesting to consider that even a low ratio does not matter if the fault lies near 100 per cent route length, i.e. 50 per cent loop length, where the healthy and faulty cores are one and the same!

Where the 'faulty' IR reading is not too far removed from the 'healthy' reading, the 'healthy' core is probably impaired at the same place as the faulty core. It may be that burn-down is being attempted to produce a difference on a PE/TDR trace. This often fails on wet telecommunications cables. The cores simply dry out and the IR reading goes up instead of down. When this happens, a useful trick to try is to rig for a bridge type test using a dried-out core as a healthy core. This must be done quickly as the core will soon become wet again and revert to a low IR value.

Severe crimping of a cable can lead to shorts which can then be located as already indicated. However, a degree of crimping can occur which does

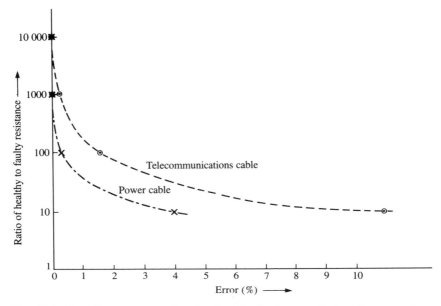

Fig. 10.3 Location accuracy related to healthy/faulty core fault resistance ratio

not reduce the IR value but does reduce transmission quality. Because this alters core-to-core or core-to-sheath disposition and increases the shunt capacitance it produces a change of characteristic impedance which can be seen by a PE/TDR, as shown in Fig.10.4. The aforementioned cautions regarding the attenuation of a transmitted pulse are equally relevant here where the mismatch being looked for is a minor one.

Cross-talk can result from coupling between two speech circuits (pairs), as shown in Fig. 3.36 and the related text, or from core transposition (splits) between pairs (see Fig.1.5 and the related text).

It may be possible to treat coupling between one or both cores of one pair and one or both of another as a straightforward contact fault. It will more likely be necessary to use a dedicated telecommunications PE set/TDR with a cross-talk location facility. This means that the outgoing pulse is transmitted down one pair and picked up on another, giving a trace feature like the one in Fig.10.5.

Fig. 10.4 Trace produced by crimping fault

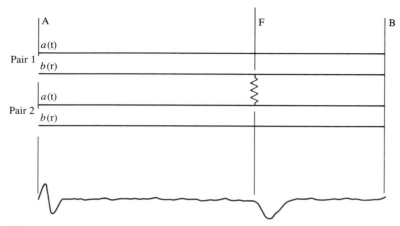

Fig. 10.5 PE/TDR trace for cross-talk location

The location of split and re-split positions (the joint where a core from one pair has been mistakenly jointed to a core from another pair, and the joint where this has been corrected) can be found with a PE set/TDR, a typical trace being that shown in Fig.10.6.

The first feature representing the core split position can be positive or negative and should be of reasonable magnitude if it is not too far from the test end as already explained. The re-split feature has the opposite polarity but will be of a lower magnitude because not only has some pulse energy been lost in the split reflection but it is by definition farther away. This being so, it may be difficult or impossible to recognize.

In any case, the joint containing the split will be located and opened up,

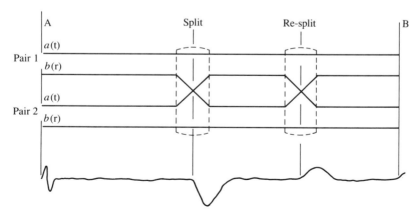

Fig. 10.6 The PE set/TDR can locate split and re-split positions

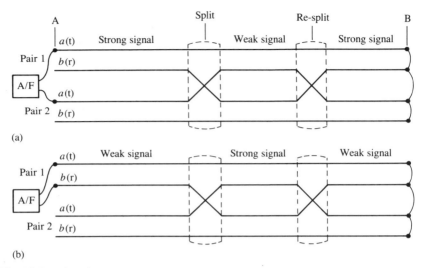

Fig. 10.7 Checking core transposition with audio frequency

so a further PE/TDR test can easily be made from this location. According to how many joints intervene, the re-split is the first or one of the first mismatches along the run and will be much easier to identify. If the cable is accessible, i.e. not direct buried but able to be picked out in cable trays or manholes, then it will be possible to find the joints in which the split and re-split have been made by applying a tone.

The transmitter connections are made to the tips (*a* legs) of the two pairs concerned, as shown in Fig. 10.7a or to tip (*a*) and ring (*b*) of pair 1 (Fig. 10.7b), in both cases with all cores in these pairs strapped at the remote end. In each configuration a strong signal will be picked up by a search coil placed on the cable on the stretches where the current carrying cores are separated, being out of their respective pairs. This is because the self-cancelling of the *go* and *return* currents is less than if they were twisted together. The opposite is true where the two current carrying cores are twisted together as a pair.

A good practical approach is to locate the split with a PE set/TDR or split locator and, after opening the joint and confirming the split, pre-locate again by PE/TDR (if available) and chase down the re-split by toning.

With regard to ground contact faults on telecommunications cables, which are also referred to in Fig.1.2a and the associated text, everything about methods and equipment has been covered in Chapter 4, 'pin-pointing', but it needs to be said that not all PTT/telecommunications companies/authorities employ or consider employing the pool of potential approach using high voltage. Those that countenance modifying the faults with a small high voltage burn-down set can do so, as this also acts as the voltage source. The equipment normally used is the type of audio frequency gear depicted in Fig. 4.15a, b and c.

It should be re-emphasized here that the fault must be a metal line in contact with the mass of earth, i.e. a core of an unsheathed cable or the sheath/armour of a cable that can be disconnected from earth at all terminations.

On telecommunications cables the happy situation often arises where there are also other faults at the ground contact point such as breaks, shorts or wet cable. A combined approach using several methods can therefore be made.

One type of fault that occurs only on telecommunications systems is 'foreign battery'. This is the appearance, on an otherwise isolated core, of a d.c. voltage of up to -48 V, derived from the exchange/central office battery voltage present on a working pair which has been impressed on the faulty core via the fault path. The full voltage appears when the fault is a metal-to-metal short and correspondingly lower voltages are observed when they are picked up through a resistive fault path.

The choice of fault location method to apply in these circumstances can be easy if there are many other combinations of fault type at the fault, e.g. other voltage free breaks or shorts. It is less so when this is the only manifestation.

In this case a resistance bridge or hand-held tester can be used if its source voltage is greater than the foreign battery voltage. The situation can be put into perspective if the foreign battery voltage is considered to be in series with the fault, i.e. in the fault path. Considering Fig.3.7, if the source voltage, V, is greater than a voltage inserted in R_f, then the test voltage has 'bucked' the interference voltage, test current flows and a balance or location is obtained.

An inverted Murray loop resistance bridge or high resistance fault locator is not applying a voltage to the fault when making a test. The foreign battery voltage or interference voltage appears in the 'measuring path', i.e. in Fig. 3.19, for instance, the voltage sensor connection from the detector D through the fault path to the cable. An inverted bridge cannot handle this. The more modern RES FT2 in Fig. 3.21 incorporates automatic polarity reversal and averaging and can cancel out lower foreign battery voltages but not those as high as -48 V.

METHODS AND INSTRUMENTS – CURRENT PRACTICE

Types of fault on metallic telecommunications cables and the basic methods used to find them have not changed significantly over the years. Cables,

258 UNDERGROUND CABLE FAULT LOCATION

materials, jointing methods and connector technology have certainly changed. These changes have had some effect on the fault type but have mainly modified fault incidence and fault management and shifted the emphasis on method. Examples of these are the move from paper to plastic insulation, the introduction of jelly-filled cables, the moves into and out of aluminium conductors and the introduction of insulation displacement connectors (IDCs).

However, basic methods have remained:

- Resistance bridge* for contact faults
- Kick test/capacity bridge (declining) for open circuits
- PE/TDR for open circuits, joints, splits, coupling and moisture as well as shorts
- Tone frequency for tracing, identification and the pin-pointing of shorts, joints and splits

Note that halving, the practice of opening joints to cut cores and test each way in order to sectionalize the faulty stretch, is frowned upon as remaking joints degrades the system. In practice it is occasionally necessary to disconnect cores to facilitate testing but this should be kept to a minimum. The practice is well known and does not involve instruments so does not require further comment.

Faults still manifest themselves in the same ways as they always did and our approach to them is basically the same. So what *has* changed? The answer to this is *technology*.

Some of the changes brought about by better technology in recent years are:

1. In resistance bridges and 'bridge type' locators:
   - Increased meter/detector sensitivity
   - Finer measurement, control and comparison of electrical quantities
   - The introduction of self-checking and averaging
   - The use of microprocessors
   - Reduction in size and cost
   - Improved reliability
2. In PE/TDR technology:
   - Use of shorter transmitted pulse
   - Better CRO displays – introduction of alpha numerics on screen
   - Greater accuracy
   - More accurate and convenient measurement techniques

---

* This heading includes all the more modern cable fault locators that still use the external cable circuit configuration required by a resistance bridge: one or more healthy cores and a remote strap.

TELECOMMUNICATIONS, INFORMATION AND CONTROL SYSTEMS

- Better presentation of results
- Use of microprocessors
- Introduction of memories
- Digitization of trace information, trace comparison and the mathematical addition and subtraction of traces
- Menu driven tests
- Recall of parameter settings
- Membrane keypads
- Use of small components and modern assembly technology to produce smaller and cheaper sets

3 In tone frequency apparatus:
- The move away from L/C tone generating circuits to quartz based and digital circuitry
- The improvement in controls, sockets and switches, etc.
- Substitution of analogue meters by digital displays and pseudo-analogue bar type displays
- Reduction in power requirements brought about by the use of smaller, more economical components and better technology
- Consequent reduction in size

Turning to the actual instruments that are currently available, the choice is wide. The types and groups of instruments resolve themselves quite clearly into the following groups:

Group 1. 'Bridge' type instruments
- Refined resistance bridges
- Comprehensive high tech locators for 'bridge type' application
- Small high tech locators

Group 2. Pulse echo sets/time domain reflectometers
- High tech PE sets/TDRs with every facility
- High tech portable PE sets/TDRs with all pair comparison and cross-talk features and memory facility
- Small, sometimes hand-held sets, usually with LCD

Group 3. Tone frequency equipment
- Complete sets of audio frequency equipment with every accessory, for tracing, identification and the location of joints, shorts, splits and ground contact faults
- Dedicated sets for tracing, identification, short and split location
- As above, but with high voltage fault conditioning
- Equipment for ground contact faults

Group 4. Routine test instruments
- The basic diagnostic test instruments such as volt/ohmmeters

In general, and this applies across all these groups except group 4, there is a great divide regarding the types of instrument used and who is using them. The upper stratum consists of the very refined and often quite expensive instruments used by the well-qualified engineer who is familiar with their operation and who takes on the very difficult situations. The lower stratum comprises basic diagnostic instruments such as volt/ohm-meters and small tone sets in daily use by faultmen.

When viewed internationally, the interface between these two strata is a grey area which also shifts according to practice in any given country. For instance, in one case, a person who is at home with a refined bridge or PE set/TDR may also carry out the sectionalizing and diagnosis. In another, a faultman will be happily and effectively using a bridge or PE set/TDR.

Currently, there is a welcome osmosis taking place downwards from the upper, through the interface, to the lower. On the ground this manifests itself as the increasing availability and use of the small hand-held PE sets/TDRs and 'pseudo bridge' devices.

This is the direct result of the influence both of advanced technology and market forces. The international communications market is huge and provides the motor for the research and development and manufacture of these units. This engine also drives the prices down and further concentrates the manufacturers' efforts in producing small, rugged, high tech instruments with no nonsense controls and displays for daily knock-about use on the network. The following examples of actual instruments are meant to be just that – examples. They are not necessarily any better than competing products with similar specifications.

In group 1, an example of a refined resistance bridge is Model 18 C, one version of which is shown in Fig. 3.8. It is standard issue in British Telecom and is in use in many PTT/telecommunications companies and authorities world-wide. It started life over 20 years ago as the British Post Office 18A designed by G.W. Crosby and combines most of the telecommunications engineer's test requirements in one accurate, well-engineered instrument. (The modern version does not incorporate the capacity bridge for breaks.) No calculations are required. A preliminary calibration procedure carried out before the actual balance means that, at balance, the final result is in ohms to the fault. There is no digital readout. The bridge is balanced by decade switching and analogue meter deflection. Other features include IR testing at 500 V nominal and 95 V nominal.

Two comprehensive high tech fault locators are shown in Figs 10.8 and 10.9. The 955 M combination fault locator from 3M Dynatel Systems Division (Fig. 10.8) is a rugged portable instrument for the testing and location of all types of telecommunications fault. It is fully computerized and can measure resistance, a.c. and d.c. voltages and d.c. current (for

TELECOMMUNICATIONS, INFORMATION AND CONTROL SYSTEMS 261

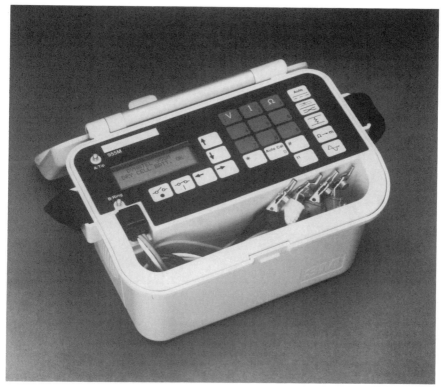

Fig. 10.8 The 955M combination fault locator (*source:* 3M Dynatel Systems Division, Austin, Texas, USA)

measuring the current in the customer loop). The input keypad and function selection switches are membrane pads. An LCD shows parameters and results. Contact faults with resistances up to 30 megohms can be located, as well as opens and splits. The tone frequencies available are 480 Hz for identification and 800 or 1020 Hz for transmission testing.

The Bartec 10T from Seba Dynatronic GmbH (Fig. 10.9) is a fully automatic computerized instrument with facilities for self-test, connection check, fault resistance test and contact fault location. The operator does not need to effect a balance, simply keying in the test parameters (conductor diameter and temperature) when requested to do so by the programme. All instructions and results are displayed on an LCD. Faults up to 10 megohms can be located and the unit can also compute the location on cables with up to four sections of different gauge conductors, the length, size and temperature of each section having been fed in beforehand.

One of several small hand-held locators available is shown in Fig. 3.10.

262 UNDERGROUND CABLE FAULT LOCATION

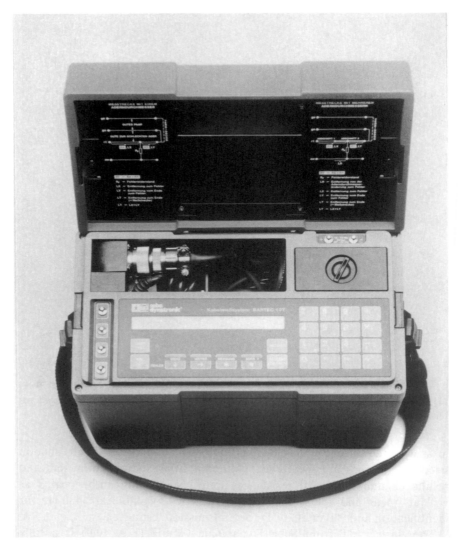

Fig. 10.9 The Bartec 10T (*source:* Seba Dynatronic GmbH, Germany)

The CFL1 from Avo International is a no-nonsense instrument with one button for checking line length and distance to fault. These are displayed on an LCD directly in ohms. As in bridge tests, one healthy core and a far end strap are required.

In Group 2, PE sets/TDRs, an example of a high tech, comprehensive set for PTT/telecommunications work is the Digiflex T12/1, T12/2 from

TELECOMMUNICATIONS, INFORMATION AND CONTROL SYSTEMS

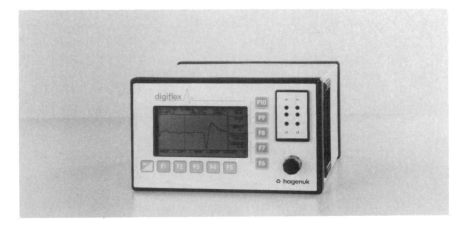

Fig. 10.10 The Digiflex (*source:* Hagenuk GmbH, Germany)

Hagenuk GmbH, shown in Fig. 10.10. This new, completely digital PE/TDR unit is small and easy to operate and has many advanced functions. The display is a large high contrast LCD. It is powered by a rechargeable nickel cadmium battery or mains or an external d.c. source such as a car battery. All control, measurement and setting functions are by key switches and one knob.

The range is from 500 m to 10 km (1 k ft to 20 k ft) and the operating modes are direct pulse echo, difference between traces, alternating traces and cross-talk measurement. The advanced version T12/2 has a bigger memory which allows signal averaging, the memorizing of 10 displays, a choice of pulse width on all ranges and the dumping of data to a PC. The T12/1 can store two displays. Both versions store setting information with display memory.

Biccotest produce the T535 rugged mid range high tech unit for portable everyday use, shown in Fig. 10.11. It is a microprocessor controlled instrument with membrane keypad controls but is one that has retained the CRO as the means of trace display. A separate LCD displays all set parameters as well as results in metres, feet or microseconds. The simultaneous display is possible of the traces from two pairs, the difference between them, crosstalk and one trace with an already memorized one, and their difference. There is pulse width adjustment and an interface for an X/Y plotter.

The T510, also from Biccotest, is shown in Fig. 3.37b. This is a very small hand-held set with LCD for the display of all settings, parameters, trace, cursor and results. All controls and settings are effected by a few simple push buttons.

264 UNDERGROUND CABLE FAULT LOCATION

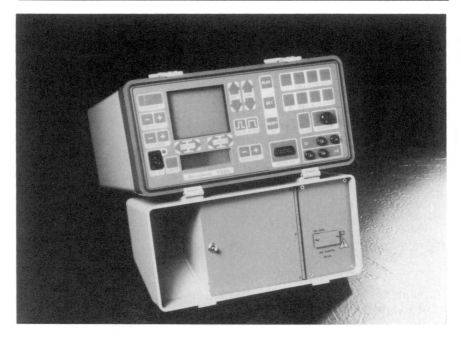

Fig. 10.11 The T535 (*source:* Biccotest Ltd, UK)

This, and the short range version, the T511, are being used increasingly by faultmen in several countries. They tend to have reservations at first but, after some training and familiarizing, come to accept this type of 'tool bag' instrument which will locate a percentage of contact faults and line taps and almost all breaks, particularly those near the customer drop. 'Looking back' down the circuit from the customer's termination has been found to be a very effective ploy by people using this instrument. This corresponds exactly with the experience in the electrical power industry mentioned in the methods available in Sec. 9.1, 'faults on low voltage systems – PE/TDR'.

In group 3, tone/audio frequency, there is a very wide choice of equipment available and the potential user should consider the test equipment requirements very carefully before making a decision.

In PTT/telecommunications work not many of the high power sets complete with all accessories are purchased. The lower powered version of these with or without accessories are in more common use. Examples of these are shown in Fig.4.15c (excluding the frame) and Fig.10.12, this latter being of a much used design with the transmitter and receiver forming two halves of a box for ease of packing up and carrying. The unit shown is the 480B split box pipe and cable locator from Metrotech. This is an up to date

TELECOMMUNICATIONS, INFORMATION AND CONTROL SYSTEMS

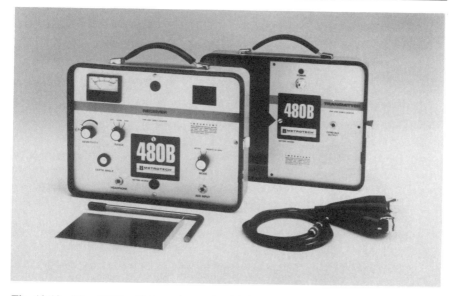

Fig. 10.12 The 480B split box pipe and cable locator (*source:* Metrotech, Mountain View, California, USA)

instrument featuring crystal control and capable of performing power line tracing, depth measurement, blind search, ground survey and the detection of buried metal objects. This type of unit can be configured as shown in Fig. 6.17.

There are several dedicated audio frequency sets for tracing telecommunications cables, the identification of cores/pairs and the location of breaks and splits. One such is shown in Fig.4.15a (excluding the frame). The techniques used in ground contact fault location are fully described in Sec. 4.5, 'audio frequency – pool of potential'. The main item of equipment required is the 'A' frame. This may or may not have the receiver mounted on it but, in any case, this item cannot easily be considered in isolation as, in most cases, it comes as an accessory with audio frequency generator/receiver sets.

The 3M Dynatel 573 D (Fig. 4.15a) has frequencies of 8, 33, 83 and 200 kHz. All controls are by membrane keypads and the LCD shows numerics, symbols and volume indication by multisegment bar graph.

A recent development in tone frequency work is the use of pulsed high voltage applied to the faulty core which makes a high resistance fault arc over and allows the audio frequency current to flow for the duration of the arc. The instrument referred to is the **TONEARC MODEL 5** from Tempo Research. It is illustrated in Fig. 10.13.

266 UNDERGROUND CABLE FAULT LOCATION

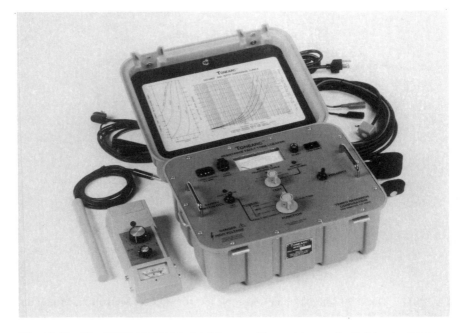

Fig. 10.13 The TONEARC Model 5 (*source:* Tempo Research, California, USA)

The peak voltages used are 300 V at the low setting and 1200 V at the medium and high settings, and the duration of the pulse is too short to damage any terminal equipment. Furthermore, the maximum currents are low – 50 mA in low and medium and 60 mA in high – so that conductors are not welded and local heating is minimal. The tone heard is 987 Hz and the pulse rate is 4/second in low and 1/second in medium and high. An approximate pre-location can also be derived from the meter reading while pulsing. This is based on the measurement of resistance to fault during the arcing period. Pin-pointing is carried out by going to access points on the route and checking for the presence or absence of tone with a pick-up coil placed directly on the cable. The tone ceases or its level reduces significantly beyond the fault. Splits and re-splits can be located by toning in the normal way and checking at joints.

The instrument is also equipped with voltage and resistance measurement features for fault diagnosis and produces a tag tone of 987 Hz, pulsed at 8 Hz, for conductor identification and a tone of 5.2 kHz for cable tracing.

Regarding group 4, every company and authority has its standard issue of volt/ohmmeter. The operators in the field are familiar with their use in diagnosis and detailed comment is unnecessary.

TELECOMMUNICATIONS, INFORMATION AND CONTROL SYSTEMS   267

10.2   Control and information systems

AUTHORITIES USING CONTROL AND INFORMATION SYSTEMS

As well as the power and telecommunications cables already discussed, there exists a multitude of cables of many different types which are used for control and monitoring, information or signalling by many authorities and companies. These are typically:

- Electricity companies and authorities
- Gas companies and authorities
- Water companies and authorities
- Railways
- Mines
- Oil and gas industries
- Petrochemical industry
- Docks and harbour boards
- Airports
- Heavy industry
- etc.

TYPES OF CABLE, VOLTAGES AND ROUTING

There are many different types of cable in use and it would be difficult to itemize them properly. For this reason communications cables have not been included in Chapter 14, Sec. 14.1. They can, however, be split into three main categories:

- Control and monitoring
- Telecommunications
- Local area networks (LANs)

The cables in the first group carry low voltage a.c. and d.c. and can be direct buried or laid in troughing, in cable trays, along walls or within wiring cabinets and control gear. There are very many relatively short routes within buildings, between buildings and outdoors from buildings to marshalling kiosks, cabinets, pillars and wall boxes.

Although all normal fault location methods as described herein can be used, the major problem confronting the troubleshooting engineer on these systems is the basic requirement of diagnosing and sectionalizing on complex networks. They comprise large multicore cables dividing and subdividing into smaller and smaller cables and wires as the route progresses to the outer limits of the circuit. There can be many stages of tee branches on tee branches also. A method that is specifically designed to solve this complex problem is described on page 271.

The telecommunications group is mainly made up of normal telecommunications cables laid with electricity power cables and gas and water pipe routes. These are usually multicore polyethylene telecommunications cables, some just with twisted pairs and some with cores as well as pairs. On power systems the cores, and often some of the pairs, are protection and circuit breaker intertripping links. Cores can also carry low voltage supplies. The main use for these cables, however, is for telecommunications, remote control and telemetering.

Carrying the above vital supplies as they do, the security of these cables is very important. They are therefore usually armoured and of larger conductor size than pure telecommunications cables and, most importantly, they are laid along routes owned by the authority or company concerned or, at least, alongside the main cable track or pipeline.

Local area networks (LANs) are communications networks, usually within one building, which provide the highway for a number of terminals (computers, printers, dumb terminals), along which information can be sent and received. The trunk cable used is coaxial with polyethylene insulation, aluminium polyester screen and braided copper screen overall. Its characteristic impedance is 50 ohms and velocity of propagation ($V/2$) 117 m/μs.

Transceivers can be connected by cables with individually screened twisted pairs and an overall braided copper wire screen. This type of cable has a characteristic impedance of 78 ohms and a $V/2$ of 100 m/μs. Compared with most networks discussed herein these cable runs are short, being tens of metres and, occasionally, a few hundred metres long.

With regard to fault type, faults other than breaks and shorts, i.e. high resistance, ingress of moisture, ground contact, etc., do not normally occur. Troubleshooting is therefore more a question of locating straightforward breaks and shorts or sorting out routes and identifying wires, as stated in 'current practice'.

CURRENT PRACTICE

In the general grouping of these special cable systems already established (control and monitoring, telecommunications and LANs), most faults can be found by one or more of the methods already described. It is the intention, therefore, in 'current practice' to give a few examples of techniques or approaches that are related particularly to peculiarities of these systems which may be helpful.

For instance, notwithstanding the current tendency to the greater use of unshielded twisted pair cable, the main feature of LAN cabling described above are short runs and dry high quality cables. The LAN itself is a refined communications system and some equipments can carry out diagnostic self-testing (which also includes what is effectively a PE/TDR test). However, the main routine troubleshooting requirements are for route and termination checking and the location of breaks and shorts, as stated earlier.

The most useful instrument for this is a low cost, short range PE set/TDR. Such instruments have a very short transmitted pulse and can resolve trouble, or the end of a cable, to within a fraction of a metre. It is best used as a 'tool bag' diagnostic instrument alongside the volt/ohmmeter. For example, it is easy to check internal cable runs for length and confirm where they go to by 'flashing' the end and watching the trace.

Regarding private telecommunications cables, the main advantages are that the route is usually well known and easy of access, e.g. a railway track or gas pipe route, and the fact that the operator normally has more freedom in being able to isolate cables and use several methods to locate a fault.

An odd situation can arise on new power cable routes whereby a telecommunications cable is laid in the same trench as the power cable as a matter of course, whether or not communications are required on that route at that time. It is relatively cheap and convenient to do this and the cable can be dug up and jointed into the system at a later date. However, because it does not need to be commissioned at the time of laying, it is very easy to 'lose the ends'. This happens often. An end is sealed, coiled up and buried unrecorded inside or outside a substation boundary and its location subsequently forgotten.

Even over several kilometres it is quite possible to trace such a cable with an audio frequency signal applied between the cores and sheath. This signal will be good almost to the end, but will decay rapidly and disappear about 50 metres from the end. In this circumstance a useful trick is to apply a voltage of several kilovolts between the bunched conductors and the screen/sheath. Whether the buried end is open and damp or properly sealed with a shrink-on end cap, it is usually easy to create a breakdown at this end. A surge generator, such as that shown in Fig. 4.3b, which operates at voltages less than 5 kV can then be used to 'bang' the end. It is then a straightforward matter to listen in the suspect area with the acoustic detector and pin-point the forgotten end exactly.

Coming now to the cable group 'control and monitoring', as stated earlier, the main problem in the field is sectionalizing and hunting down an earth fault. The presence of this earth fault will probably have been indicated initially by earth fault monitoring equipment, for instance on an unearthed d.c. supply system. Typical of such a problem is the situation shown in Fig. 10.14.

This is a 110 V d.c. supply system from a large power station or, substation battery. The positive and negative poles are fed to a multitude of

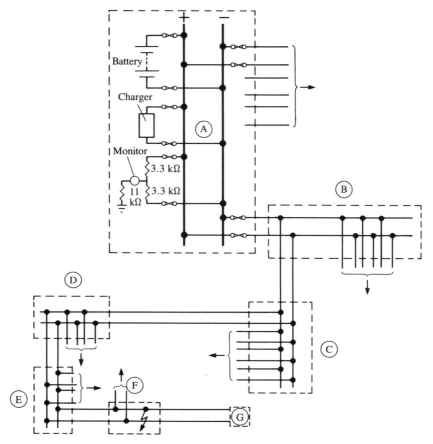

Fig. 10.14 Earth fault on complex d.c. system

destinations such as protection relays, circuit breaker trip and closing coils, etc., via a multiteed system routed from building to building, building to kiosk, wiring cabinet and wall box. In Fig. 10.14, the fault is shown in a small terminal box which could well be outdoors in a compound and have become damaged, allowing ingress of moisture and subsequent corrosion.

High impedance earth fault monitoring detects faults up to 25 kilohms and a local indication and remote alarm are initiated. An easy way of chasing the fault down is to momentarily disconnect circuits until the indication disappears and repeat this process down through the network until the fault is found. This is never normally done. Such is the importance of these auxiliary supplies that they are never disconnected in such a wholesale manner.

TELECOMMUNICATIONS, INFORMATION AND CONTROL SYSTEMS

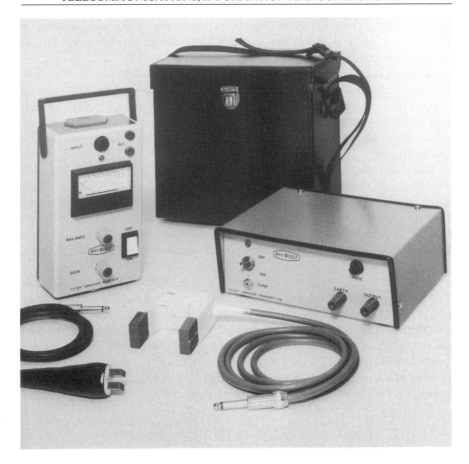

Fig. 10.15 The T273 'Grouser' (*source:* Biccotest Ltd, UK)

The equipment used to trace the fault through the system is a low frequency system such as the T273 'Grouser' shown in Fig. 10.15. It comprises a transmitter, T273/T, a receiver, T273/R and small, medium and large pick-up coils with U-shaped cores. An extra large one is available as an option.

The transmitter produces a low frequency signal of 11/12 Hz which is injected between a live d.c. conductor and earth. Such a low frequency is required in order to ensure that the signal is drawn by the resistive path (the fault) and not by the system capacitance and to ensure there is no capacitive carry-over after the fault. Even so, there can be some capacitive current on large systems that can have a total capacitance of tens of microfarads.

To carry out a fault location the transmitter is connected to the faulty pole at source (location A in Fig. 10.14) and one of the pick up coils is placed over each outgoing leg in turn. The circuit showing the largest deflection is the one feeding the fault.

Occasionally several circuits manifest significant deflections. This can happen if the fault is a high resistance one and the system has a lot of capacitance. The reference terminals are then connected to the faulty pole and earth to pick up a reference signal. Turning the balance control produces a null point on a healthy circuit. It is not possible to get a null on the faulty one.

When the correct (faulty) outgoing leg has been identified the signal is confirmed on the incomer at location B and the outgoing legs at B checked until the signal is found on one of them. In like manner the signal is traced into and out of each location, C, D, and E, and found to enter but not leave F. By this time the fault will be evident in box F and cleared. The confirmation of this will be the cessation of the alarm signal in the negative pole back at source.

Note that on any unearthed system it is essential to locate and clear the first earth fault indicated very quickly. All the time it stands on the system the statistical chances of another earth fault occurring in the other pole increase. If this happens a 'cross-country' fault occurs, shorting the two poles via the earth path. Furthermore, another fault on the same pole can cause a trip or closing coil on one circuit breaker to be energized by the closing of a trip or close contact on another circuit breaker, as a feed can be picked up via the two faults and the earth path.

11

Under-floor and under-road, pipe and soil heating systems

11.1 Types of system

Electrical surface heating is used in buildings, roads, ramps, playing fields, running tracks and horticultural establishments. There are also very many pipe heating installations on industrial sites where a heating cable is run along or wrapped around a pipe in order to maintain the product within the pipe in a mobile state or to prevent freezing.

All types of cable used except one have resistive metal alloy conductors and run from normal low voltage a.c. mains at between 110 and 415 V. Their size and length are chosen to dissipate a certain wattage for a particular heating requirement.

Most resistive heating cables are single-core with XLPE insulation, sometimes with a heat resistant PVC sheath. Some also have a galvanized steel or copper wire braid screen and PVC sheath.

The other main group is made up of resistive heating cables with mineral insulation and copper screen. These are essentially the same as the well-known mineral insulated distribution cables used world-wide in hostile environments. The insulation is highly compacted magnesium oxide.

The exception is a type of flat heating cable with two normal copper conductors but with a semiconducting dielectric between them. They are screened with metal braid and and plastic covered. This type of dielectric heating cable is mostly used as pipe heating, as is the mineral insulated cable previously described. If a resistive heating cable touches another a 'hot spot' is created which quickly becomes a burn-out. However, it is permissible to touch together, or cross over, the legs of dielectric heating cable as the molecular structure of the heating element or 'core' changes at the point of contact and the resistivity is modified to prevent a hot spot developing. In other words, it is totally self-regulating.

Normally, the maximum running temperature for heating cables is 80°C. Under-floor heating cables are usually laid in 'mats', i.e. discrete lengths laid out in a zigzag pattern and covering a chosen area of floor. The two ends are usually brought out close together as 'cold tails' about 3 to 4 metres long. These are tails of normal cable jointed on to the heating cable. The joints are often factory made.

The cable should be very carefully laid with fixed spacing between runs. This can be achieved by running the cable through slots in special plastic spacer bars. No leg should touch or be close to another. The wires are embedded in a uniform sand and cement screed. It is during screeding that they are most vulnerable to damage. Below this screed comes the concrete floor slab and, below that, thermal insulation and finally a damp-proof membrane.

Soil heating wires can also be laid in zigzag pattern mats or, as in the case of running tracks, side by side in (physically) parallel runs.

The foregoing is not a complete detailed description of all types of installation but is hopefully adequate to illustrate how the type of cable and installation decide the approach to fault location and the choice of method.

11.2 Choice of method

The two major factors to be considered are whether the cable is unipolar or bipolar and whether the *exact* route is known.

A unipolar system, i.e. a cable with no metal screen or second conductor, can have an earth on it or a break in it. The earth fault would be where the insulation is damaged and the conductor is in contact with the screed.

A break could be a clean break well insulated from earth on both sides, but would more probably be in contact with the screed on one or both sides of the break. It is also likely that there would be a finite resistance across the ends of the break as a result of burning.

There are several possibilities for finding these faults. The earth fault can easily be pre-located with a bridge type instrument. Both ends are available at one place and a location will be very accurate. However, it is of little use knowing that the fault lies at 26.23 per cent route length if neither the route nor its length are known exactly!

Plans of the layout of and connections to heating mats are available at the time of laying, but it is unusual for them to be at hand some years later when a fault occurs. It is also certainly possible to trace out a cable run using audio frequency equipment such as that already described, but extremely difficult in practice. The legs are close together, making it necessary to use a small pick-up coil and low signal strength, and the

UNDER-FLOOR AND ROAD, PIPE AND SOIL HEATING SYSTEMS 275

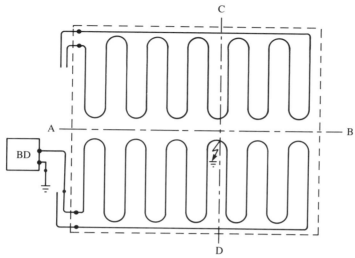

Fig. 11.1 Underfloor heating cable fault

routing of the 'far end return leg' is often along a wall skirting adjacent to the turn-round points of the other legs.

In the absence of plans it is very difficult to say exactly how a floor heating section is set out, i.e. just how the zigzag legs run. A layout might be encountered such as that shown in Fig. 11.1. If the route is not *absolutely certain* a pre-location of x per cent could be taken as being up one zig when it is actually down another zag and at the other end of a room! It is most unusual for pre-location information to be sound enough to justify chiselling out the screed and exposing the cable. To make matters worse, the floor is often covered by fancy tiles or wooden parquet blocks.

PE sets/TDRs cannot be used on unipolar systems. If there is ground contact a much more promising approach is the pool of potential method using HV d.c. as described in Chapter 4. The HV d.c. source should be one of the small sets limited to around 2 kV. A pulsed test voltage of less than 1 kV is usually sufficient. The ground spikes normally used have to be dispensed with and a small pair substituted. These can be made from small screwdrivers sharpened at the ends, connected to the galvanometer leads and insulated down to the tips.

The fact that, in under-floor heating fault location, the run of the cable is not being followed is a perfect illustration of the anatomy of this method. As emphasized in Chapter 4, the return currents spread out radially from the fault in the mass of earth (floor screed) and the cable run is totally irrelevant – a distinct advantage in under-floor heating fault location.

Referring again to Fig. 11.1, sampling voltages with the small spikes along a line AB along the centre of the room shows a minimum where the line CD intersects. A subsequent traverse along this perpendicular line CD will soon indicate the steep voltage gradients around the fault.

There are, however, problems with small areas. On a long external cable route the currents emanating from the fault return to the HV source at its ground connection a long way away. In a confined space there is always the possibility that other 'pools' of potential occur where there are earthed objects such as a pipe. Therefore the whole room should be checked and the 'pool' with the severest voltage gradients will normally be over the fault. Note that it often helps to dampen the floor where the small spikes are pressed into the surface. They can also be carefully pushed down the gaps between corners of tiles without causing damage.

Bipolar installations are a different matter. All pre-location methods are possible including PE/TDR. Pin-pointing by the pool of potential method may be applicable if the test is between the sheath and the mass of earth (floor). Audio frequency pin-pointing can sometimes be used but the fault needs to be of very low resistance for this to be effective.

11.3 Practical approach

UNDER-FLOOR AND UNDER-ROAD INSTALLATIONS

It should be clear from the foregoing that, in under-surface fault location, pre-location is almost always a non-starter and that a pin-pointing method has much more chance of success. As always in fault location, the peculiar electrical and practical aspects of a cable or system should be exploited to the utmost.

The facts are:

1 The cable is short.
2 It may be sheathed or not.
3 It has a significant resistance end to end.
4 The fault will normally have resulted from initial damage which has caused a burn-out.
5 Both ends are easily accessible.
6 The area within which the fault lies is small.

After the usual careful diagnosis, despite what has just been said about pre-location, a pre-location should be made. This is just in case the fault lies very close to one end, in which case it may easily be found at or near one of the cold tail joints. Also, the operator may be absolutely certain of the layout of the mat from plans or tracing.

If, as is more likely, the fault is not near an end and the route is not

UNDER-FLOOR AND ROAD, PIPE AND SOIL HEATING SYSTEMS 277

known, the modified pool of potential method described above may be used for pin-pointing. However, there are more effective methods of pin-pointing than this if certain equipment is or is made available. These depend on temperature detection.

Whether the fault is a contact fault or a break it usually exhibits a leakage path to earth or screen, and if this path is stressed with the small burn-down set a current can be made to flow. It does not particularly matter whether the path is conductor to ground, conductor to screen, conductor to conductor or even the gap between the ends of a break. The intention is not to burn down the fault but to keep current flowing for five or ten minutes in order to create a *hot spot* on the floor or road surface. This much is reasonably straightforward. The difficulty lies in having suitable equipment on site to detect the hot spot.

Bolometers are in common use. These are contact type temperature sensors that can traverse the surface.

There are many other instruments that do this in an easy, indeed spectacular, way. These are thermal imagers. If the surface is panned with a sophisticated thermal imager and the burn-down set switched on, after a few minutes the hot spot over the fault can be seen to be growing. It is then very easy to confirm the location and chisel out the small area over the fault.

The thermal imager is, however, far too expensive to justify purchase for this purpose. Even the cheapest versions are too expensive. At the other end of the cost scale, the palms of the hands are remarkably sensitive and selective in this application. In between these extremes lies the best solution, which is the basic type of thermal 'gun', i.e. a hand-held temperature sensor which displays the target area temperature digitally.

The human brain does not take kindly to logging and memorizing a sequence of changing numbers. However, it is very good indeed at analysing patterns and detecting level changes. To capitalize on this, a detector has been made available that produces an audible tone, the frequency of which varies with the signal, i.e. with variations in temperature.

The operator simply pans the surface with the detector until a sudden localized rise in frequency identifies the hot spot. The rig is extremely sensitive and will indicate temperature changes of less than one degree Celsius. In practice, a hot spot with a temperature of several degrees higher than that of the surrounding surface is soon produced.

It is good common sense to check the area carefully *before* starting the burn-down process. There may well be warm areas due to water pipes, drains or equipment or appliances. These can be marked and their temperatures noted before switching on the current. Figure 11.2 shows such a set in use. It can also be effective as a pin-pointing device in low voltage

278 UNDERGROUND CABLE FAULT LOCATION

Fig. 11.2 Thermal detector in use

cable fault location for detecting warm areas in roads and pavements, as mentioned in Sec. 9.1.

SOIL HEATING INSTALLATIONS

The practical approach to fault location on soil heating mats is dictated by the type of heating wire and the physical layout of the mat.

The wire is normally of the non-screened type so that, in almost all cases, there is a simple ground contact situation. This, combined with the fact that on a running track or field there is overall access, means that the pool of potential method using the normal ground spikes is straightforward.

On a running track, a resistance bridge or 'bridge type' instrument can also be used. Such a test gives excellent results because the faulty cable is a loop of conductor with both ends available in one place and whose route, resistance and length are known exactly. Therefore a good pre-location can be quickly made which saves walking the whole track.

PIPE HEATING CABLES

Faults on mineral insulated cables can be pre-located by 'bridge type' instruments with auxiliary wires run out to the far end if necessary because the route lengths are small. PE/TDR can also be used if there is a low resistance fault or break. It may be thought that there is no need for great

UNDER-FLOOR AND ROAD, PIPE AND SOIL HEATING SYSTEMS

accuracy as lengths are short, but the opposite is normally true because the pipe and cable are often covered in lagging, making it impossible to confirm the location by inspection only.

A surge generator can be used for pin-pointing providing low enough voltages are employed, safety rules are observed and there are no constraints imposed by hazardous areas.

Sections of cable of the dielectric heating type are about 80 m long and instruments such as that shown in Fig. 3.21 have been used successfully for some years. The leaky dielectric 'core' is, of course, a shunt path across the fault but is uniformly distributed along the whole route.

12

Lighting cables – motorway/highway, road and airfield

12.1 Fault location philosophy

In considering fault location on a special type of cable network, the question arises once again:

'Is this just another cable with a fault on it?'

The answer, as always, is:

'Yes, but . . .'

In the reservations, provisions, difficulties and differences implicit in this *qualified* response lie the keys to successful fault location on any specialized or dedicated system.

It is the *differences* that, though they may throw up special difficulties, also offer possibilities that do not exist on other types of network. These differences should be assiduously exploited to create advantages for fault location.

Consider then the special features of lighting cable systems with a view to using them in making fault location easier:

1 Street lighting cables may be similar to power supply cables but they are *longer* for any given cross-section because they carry smaller currents and volt drops are less.
2 Most are or can be switched off during daylight hours. Note that some are still controlled by individual or group switching by time switch, but most are switched, individually or in groups, by photo cells.
3 Their 'consumers', lamp columns, are easy to access.
4 To a great extent their routes are obvious.
5 Some lamps have teed off services. In the other system the cable is

LIGHTING CABLES – MOTORWAY/HIGHWAY, ROAD AND AIRFIELD

terminated in each column and run out again, so that it is easy to sectionalize the cable into *very short lengths*.

6 They are regularly scouted and reported on.
7 The electrical integrity of cable sheaths on motorways, highways and major roads can usually be checked.
8 Some highway and all airfield lighting cables are part of high voltage constant current systems.

12.2 Types of system

LOW VOLTAGE

Most ordinary street lighting is fed by dedicated low voltage cables which can belong to the electrical supply authority or to a municipality. There are still in existence fifth-core systems whereby the street lamps are fed by the neutral and a special small cross-section core in a three-phase main cable. These therefore belong to the supply authority and any testing on or treatment of the fifth core is obviously affected by its proximity to the three main phase conductors, two of which stand at phase-to-phase potential to it. These systems will not be covered separately as the methods used are largely the same as those to be described, given the constraints outlined above.

Where dedicated public lighting cables are used, the lamp columns may be teed off a three-phase cable. Such tees will usually be single-phase. There are also stretches of single-phase cables with tees. Some cables are paper-insulated, lead covered but most are of armoured plastic construction.

In certain areas, large sections of a network or sometimes whole networks have the cables looped 'in and out' of the columns, which makes for straightforward fault location as sections can be easily isolated. Low voltage motorway cables are run out of feeder pillars fed from the distribution authority low voltage network. They are usually laid across, to and along the central reservation down which the lamp columns are situated. Two-core cables can be used, looped in and out of the columns. Otherwise, for long distances or to feed high wattage lamps, four-core cables are used. Occasionally all three phases are looped in and out of each column but, more normally, single-phase services are run in. From a point of view of safety it is not advisable to have phase-to-phase voltage present in a column in case it is damaged.

Cable construction is typically 16, 25 or 35 mm^2 PVC insulated cores with steel wire armouring and a PVC sheath overall. The 16 mm^2 cable is often looped in and out, the larger sizes being jointed into the columns. Instead of PVC, a less toxic and only slightly more expensive XLPE core insulation is now being used.

A three-core cable is also used with one core as the CPC (circuit protective conductor or earth) when the armouring is inadequate for carrying earth fault return current. The armour (CPC) is sometimes earthed at the feed point, particularly when this is requested by the supply authority, but more often by an electrode at the last or the penultimate column (refer to BS 7430 of the British Standards Institution).

HIGH VOLTAGE

Some road lighting and all airfield lighting is fed by high voltage cables. Voltages between 2 and 8 kV are used, the lamps being connected in series and each one having about 50 V across it. They are fed from a constant current transformer/regulator from which, in the case of street lighting, the HV single-core cable runs along the route to be lit and returns either back along that same route or along another, the lamps being positioned wherever necessary on it. Therefore the *go* and *return* legs may or may not be found together at any given point, depending on the design of the system.

The HV cables on an airfield usually leave the constant current transformer/regulator and run together for a distance before splitting up to form a ring on which the lamps are situated. Unless some device is employed to prevent it, it is quite clear that any one lamp burning out will break the circuit and leave all the lamps out. In fact, this is avoided in two ways.

In the case of high voltage street lighting, each lampholder has two sprung metal contacts which would short the lamp out if they were not separated by a thin insulating film. While the lamp is healthy, the volt drop across it is not enough to rupture the film. If it goes open circuit, however, the whole series circuit is interrupted and a voltage of several kilovolts appears across the film. This is easily enough to rupture it and the two sprung contacts come together and current flows in the loop again.

In all airfield lighting circuits and in some street lighting circuits, current transformers (CTs) are connected at each lamp position. Thus the primary loop is made up of the cable core with all the CT primary windings in series. Unless there is an open circuit fault in the cable or a CT primary, the HV circuit loop will remain intact. The lamps are fed by the CT secondary windings so that a lamp burning out does not affect the primary circuit at all. Both single-core unscreened and single-core screened cables are used in these high voltage lighting systems. In street lighting systems the *go* and *return* cables can often be found together in one location, but on airfield lighting systems only one cable is accessible at the lamp position CTs. These CTs can be direct buried or in pits and the older ones are permanently connected and sealed with compound. Modern ones can usually be disconnected with waterproof/water-repellent plugs and sockets.

12.3 Methods of fault location

LOW VOLTAGE CABLES

A major consideration in deciding the best fault location method is availability of equipment. The engineer or technician dealing with street lighting faults on a regular, routine basis is unlikely to have the whole range of equipment, i.e. arc reflection, impulse current, PE/TDR, burn-down set, shock wave generator, pool of potential, audio frequency equipment, etc. Indeed, some of these techniques, such as arc reflection and impulse current, are inappropriate anyway. The instruments most likely to be available are:

- Basic PE set/TDR
- Resistance bridge or 'bridge type' instrument
- Basic audio frequency gear
- Surge generator (occasionally)

Bearing in mind the comments in fault location philosophy above, the route will be known, the loads will be disconnected and access to the cable cores at many points will be easy.

If the fault is a break or a low or medium resistance short, a simple, short range PE set/TDR can be used to great effect. Even though a trace viewed from one end of the cable may be impossible to interpret, it is quite feasible to apply the set at several terminations. When the set is near to the fault a very distinct trace feature will be apparent. Figure 12.1 shows a cable with tees into lamp columns with an open circuit fault on the main between lamps 5 and 6.

Traces viewed from A or B will be too complex to interpret but with the set connected at 5 a very clear open circuit feature will be visible and an accurate distance measurement will indicate that the fault lies very close to the lamp 5 tee joint. A quick check from lamp 6 will confirm that it is between 5 and 6 and another accurate confirmatory measurement can be made. Low resistance shorts can be accurately pre-located in exactly the same way.

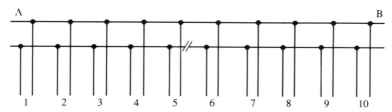

Fig. 12.1 Open circuit fault on street lighting cable

284 UNDERGROUND CABLE FAULT LOCATION

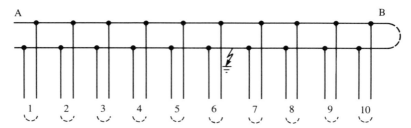

Fig. 12.2 High resistance earth fault on street lighting cable

It should be noted here that, because street lighting fault pre-location is invariably finalized over relatively small distances, accuracy is high and pinpointing is not always absolutely necessary. If a PE set/TDR is not available or cannot be used, as in the case of a high resistance fault, a high impedance fault locator such as the RES FT2 shown in Fig. 3.21 can be used most effectively both to 'home in' on the fault and to pre-locate it accurately. In Fig. 12.2 there is a high resistance earth fault just beyond the tee joint to lamp 6. A high resistance fault locator can be connected at A and an 'end strap' made between the two cores at, say, lamp 2. This will result in a reading of 100 per cent. In other words, the actual fault is effectively connected to the arbitrary 'end' at 2 by a length of core.

Similarly, a 100 per cent reading will be obtained in any test carried out with the 'end strap' put on at 3, 4, 5 or 6. However, if the strap is applied at 7, 8, 9, 10 or B, a correct percentage reading of the fault distance will be obtained. Although a measurement can be made from any of these results, it is evident that the fault lies between lamps 6 and 7, and a very accurate fix can be obtained by applying the test set at B and putting an 'end strap' on at 6.

A further ploy which gives extreme accuracy whenever an earth fault is known to lie between two lamps is to run wires overground as shown in Fig.12.3. Note that this is a four-wire test similar to that shown in Fig. 3.13 and loop length is synonymous with route length; of course, no healthy core is required so this approach can be used when all cores are faulty.

The above techniques can also be used in low voltage motorway lighting cables with the rider, however, that rapid easy access to all lamps is not a straightforward matter on busy trunk routes. Yet again, differences in a system close some doors but open others. For instance, the sheath/armouring of such cables has to be tested against the mass of earth at 3 kV d.c. for one minute on installation. A subsequent cable fault will invariably involve damage to the plastic oversheath and, because the metal sheath/armour can

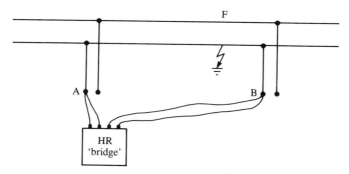

Fig. 12.3 Accurate bridge location between two lamps

usually be disconnected from earth, a pool of potential test can be carried out (see Figs 4.11 and 4.12 and the associated text). Moreover, in this situation, the directional audio frequency method outlined in 'practical approach' in Sec. 4.3. is very effective.

HIGH VOLTAGE CABLES

Troubleshooting on high voltage constant current lighting circuits is greatly facilitated if the *go* and *return* cables can be found in close proximity at several points so that they can be shorted together to check continuity or to carry out a resistance bridge test. Unfortunately, in the main area of application of such systems, i.e. airfield lighting, the two cables cannot normally be found together at access points. To add to this difficulty, the presence of the lamp CT primaries in series at many points along the route creates a circuit such as that shown in Fig. 12.4.

Even when presented with all these difficulties, the operator can exploit a circuit *difference*. This is that the start and finish points are physically close together at the constant current transformer/regulator.

'Bridge type' tests can be carried out but they require calculations to be made of the CT primary winding resistances and the keeping of careful records of percentage locations made for earth faults at various positions. This approach, then, is more theoretical than practical.

PE/TDR is of no use either. If the cable is unshielded it cannot be applied as it is a two-pole test and does not function when connected between a single conductor and ground. Even if the cable is screened, the pulse cannot travel through the first CT primary winding encountered. A positive trace feature is displayed, similar to that of an open circuit or end.

This leaves only 'direct overground' approaches such as the pool of potential and the ballistic galvanometer methods. The ballistic galvanometer approach is applicable to airfield lighting loops of unshielded single-core

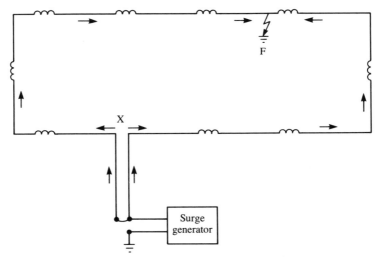

Fig. 12.4 The ballistic galvanometer – a 'direct overground' method for airfield lighting cable faults

cable. Figure 12.4 shows such a circuit with a normal surge generator connected between earth and the two cable ends commoned together. The technique works because, in this type of circuit, the *'go'* and *'return'* legs are together at the constant current regulator/transformer but are separated along the lighting route, only running together on the access stretch.

A ballistic impulse detector is required which responds to the *direction* of the impulse. As shown in Fig. 12.4, the two ends of the circuit are disconnected from the constant current transformer/regulator, commoned and connected to the surge generator output, the other end of which is connected to earth, preferably via a ground spike away from the system earth.

The surge voltage applied depends on the d.c. proof impulse rating of the cable and transformer primaries. As an example, single 8 AWG plastic insulated wire with a nominal rating of 5 kV has a d.c. proof test value of 25 kV and a basic impulse level of 75 kV. The lighting transformer primary windings might have a proof test value of, say, 15 kV. The impulse is only applied for around 100 μs every few seconds; therefore the cable and transformers specified above could withstand surges of up to 15 kV without any insulation being overstressed.

When the circuit is impulsed the signal splits two ways at X and recombines and flows to earth at F. The operator simply walks the route at a steady pace with the search coil more or less over the cable. On the approach to the fault, the centre zero meter will kick to one side with each impulse. When the fault is passed it will kick to the other side. The fault can therefore be located accurately and in a short time, as the operator can walk several kilometres of route in about half an hour.

There will, of course, be another bidirectional indication at X, but the location of this point will be known and the indication can be discounted. If required, the location can be confirmed using the acoustic detector.

It is clear from the circuit and fault configuration that the pool of potential method can also be used, but this would normally take longer.

13

Optical fibre cable systems

The following discourse on optical fibre cable systems is a brief overview of a very important communications medium as it relates to metallic cable fault location. It is not intended to be a complete and detailed description of optical fibre communications.

13.1 General

An optical fibre cable is simply a link in a communications system which works by the transmission of light, from a laser or LED source, along a glass fibre. Except for some short non-critical data links which use large diameter plastic-coated silica fibre, the fibre make-up is a glass core surrounded by glass cladding of different refractive index. The glass used is of very high purity.

Fibre optic cable has many advantages over metallic cable. It is totally immune to interference from adjacent electrical circuits and equipment and signal security is almost absolute as it is very difficult to tap into. It is also relatively cheap, small and easy to install. Attenuation is very low compared with metallic cable so that very many more voice channels can be accommodated and repeater distances are much greater. It finds use, therefore, not only in trunk and junction networks, but in hazardous areas, military applications, CATV, high voltage and heavy interference situations, etc.

13.2 Types of fibre

In a 'step index' fibre, the light travels along the core and is contained within it by total internal reflection from the junction between the core glass and the cladding glass, as shown in Fig. 13.1.

In 'multimode' systems, rays at an angle greater than the 'acceptance cone angle' pass through the core/cladding junction into the cladding and are lost. Some rays or 'modes' travel straight up the middle of the core.

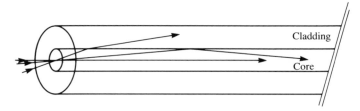

Fig. 13.1 'Step index' fibre

These are called 'low order modes'. Others, called 'high order modes', have farther to travel because they are being reflected and re-reflected off the core/cladding junction and therefore arrive later than the low order modes. This produces 'modal dispersion', making the received pulse more rounded and less distinct than the transmitted pulse. Therefore multimode step index fibres have a limitation in that they cannot be used for long distance transmission.

There is another type of multimode fibre called 'graded index' fibre which can be used for transmission over longer distances. In this, the refractive index of the core changes gradually from a high value in the centre to a lower value at the junction with the cladding. As shown in Fig. 13.2, a low order mode along the axis will travel slower because it moves in the high refractive index glass at the centre. The higher order modes, as they travel towards the periphery and are gradually 'bent' back by the changing refractive index of the glass, are travelling faster where they go through glass of a lower refractive index. Thus the arrival times of all modes are considerably evened out. This decrease in modal dispersion results in less rounding off and distortion of the received pulse and a higher bandwidth. Typical core/cladding sizes in micrometres are 50/125, 62.5/125 and 100/140.

The type of fibre used for long distance transmission is 'single mode' fibre. This is a step index fibre but with a very small diameter central core which only admits one mode. Figure 13.3 shows such a fibre with a core diameter of 8 μm and an outside diameter of 125 μm used in 1300 or 1550 nanometre transmission.

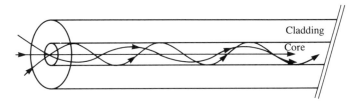

Fig. 13.2 'Graded index' fibre

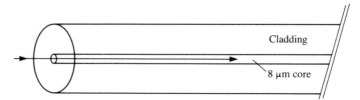

Fig. 13.3 'Single mode' fibre

13.3 Specifications and usage

Signal attenuation is low at three 'windows' of wavelength. These occur at 850, 1300 and 1550 nanometres, as shown in Fig.13.4. Therefore telecommunications trunk transmission has evolved through the use of multimode fibre at 850 and 1300 nm, requiring repeaters at 6 to 8 km and 10 to 20 km respectively, to the use of single-mode fibre at 1300 and 1550 nm which only require repeaters at much greater distances up to 100 km in terrestrial communication, with some submarine cables having even greater unrepeated lengths.

The effectiveness of an optical fibre transmission cable is determined by its losses measured in decibels.

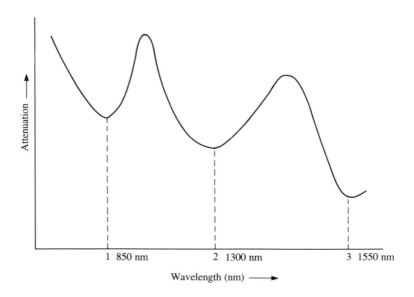

Fig. 13.4 Signal attenuation at three 'windows' of wavelength

13.4 Losses, connections and splices

Losses occur in three areas: connectors, used to connect terminal equipment at each end, and splices between sections of cable, and in the fibre itself. Despite the high purity of glass used, some losses do occur in the fibre due to backscatter of light from impurities. This is called 'Rayleigh scattering' and can vary between several decibels per kilometre in short wavelength multimode transmission to less than 1 dB/km in long wavelength single-mode transmission.

When mated, the two made off halves of a connector permit the butting of two fibre faces which have been almost perfectly ground to 90° planes. The loss involved is of the order of 1 dB. Some connectors now have angled or radiused ends to reduce losses even further.

Splicing, both indoors and in the field, is carried out using very refined fusion splicers. Even when the circular ends of the cores to be offered up together are as small as 8 μm, the machine can butt them together automatically. This is usually done by the 'profile alignment' system. The two fibre ends are carefully cut with a separate cleaving tool to produce near perfect 90° faces. These are then clamped in the splicer jaws, roughly aligned in two dimensions and checked for cleanliness and cleave angle. The machine then automatically aligns them and an arc is fired to weld them together. During the arcing, the fibres are driven together through 5 to 10 μm to make the joint. Currently, very accurate splicing is being carried out and typical splice losses are less than 0.1 dB.

When a system is designed, a so-called 'loss budget' is established for a particular link. This is made up of connector losses plus splice losses plus fibre losses plus a margin to allow for future repair splices. The budget should fall well within the maximum loss allowed before signals become too corrupted and errors occur. This figure is around 25 dB in telecommunications, but can be higher or lower for other applications, e.g. data communications, 10 to 15 dB; submarine, 30 dB.

13.5 Instruments for testing and fault location

The only proper way to measure system loss is to use an optical attenuation measuring set which comprises a light source/transmitter and a receiver/optical power meter. With these the loss between the transmitter and receiver can be measured accurately.

Reflection tests are also carried out using an optical time domain reflectometer (OTDR). This is the optical equivalent of the PE set/TDR, although the significant difference between them is that, because Rayleigh backscatter returns from every part of the fibre, the trace produced by the OTDR is a graph of distance against loss.

In optical fibre cable testing and fault location there is a direct and

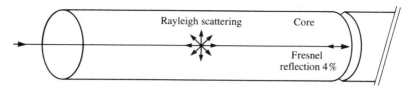

Fig. 13.5 Rayleigh scattering and Fresnel reflection

obvious comparison between an open circuit or break in a metallic cable and a break in a fibre. In both cases the break interrupts transmission and can be seen and located using PE/TDR technology.

The contact fault or earth fault in metallic cables has no direct equivalent in fibre systems, although a signal can actually leak out of a fibre core into the cladding if there is a small bend in the fibre. This is called a *micro bend*. The angle of light to the core/cladding junction on the bend allows it to be refracted out of the core instead of being reflected back into it. This phenomenon is utilized in fibre identifiers wherein small motor-driven bending heads carefully bend the fibre to extract live signal or a trace/ identification signal.

OTDRs enable the loss pattern of a fibre to be viewed and analysed by launching a laser pulse into the fibre and measuring returning reflections. Figure 13.5 shows the two phenomena which make this possible: *Rayleigh scattering* from impurities along the fibre and *Fresnel reflection* from an air/ glass interface, such as the reflection seen in a normal window pane. A clean fibre break will be shown as a large Fresnel reflection. It should be noted that only very tiny reflections are involved, e.g. only 4 per cent maximum from a 90° break, and it is difficult to view these in real time.

The OTDR, therefore, performs a continuous averaging of repeated pulses so that a trace with discernible features slowly grows on the screen. Such a trace is shown in Fig.13.6. Note that the first length of fibre has a particular gradient which can be measured from the screen as a certain attenuation in decibels per kilometre, while the second length of fibre has a steeper gradient, indicating that it has a greater attenuation.

Splice and connector losses can be measured but their values should not be taken as accurate because an OTDR is measuring reflected light, not actual loss. Furthermore, splice losses are currently so small as to be very difficult to recognize in certain cases.

The main function of an OTDR is to measure the *distance* to a particular feature. This is usually an empirical measurement with due allowance being made for the spiralling effect of the fibre within its tube in the cable, which can result in the fibre being up to 2.5 per cent longer than the actual cable.

OPTICAL FIBRE CABLE SYSTEMS 293

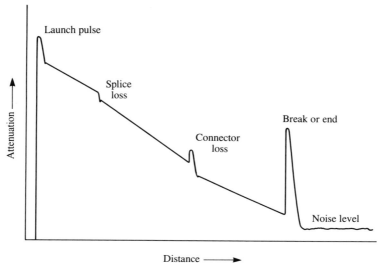

Fig. 13.6 Typical OTDR trace

PART FOUR

Cables

14

Information on cables

14.1 General

It is largely true to say that anyone dealing with cable fault location has a working knowledge of cables. There is a need, however, for a handy source of facts and figures to do with cables which relate particularly to fault location. It is hoped that the following comments and data go some way to addressing that need.

Note that telecommunications cables are not included. This is because there is such a large number of cables that vary considerably in size, make-up, pairage, insulation, screening and protection. Typical core sizes, core resistance, screening and velocities of propagation have already been discussed. Manufacturers' information should be consulted for further details.

14.2 General construction

At very high transmission voltages there is not a great call for cable information related to fault location because no routine fault location is required. As explained earlier, gas and oil pressure cables do not normally develop core faults as they are effectively policed by gas and oil pressure monitors and alarms and sheath fault testing.

At medium voltages, paper-insulated cables are still encountered. They have three copper or aluminium cores with layers of oil-impregnated-paper insulation, layers of oil-impregnated-paper belting insulation over the cores, a lead or aluminium sheath, bituminous hessian protection and steel tape or wire armouring with bituminous hessian or PVC protection overall. The cores may also be screened with a paper/aluminium foil laminate or copper or aluminium tape, such screening being necessary at voltages over 11 kV. An example is given in Fig. 14.1.

An alternative construction is three individually screened single-core

298 UNDERGROUND CABLE FAULT LOCATION

Fig. 14.1 Three-core, 150 mm², 6.35/11 kV, screened PILS cable with PVC oversheath (from *Electric Cables Handbook*, p. 285, ed., D. McAllister, Granada, London, 1982)

cables laid up together with sheath, armouring and protection overall. Such a cable is shown in Fig. 14.2.

Table 14.1 gives the minimum insulation thickness and maximum electrical stress for both types of paper-insulated cables.

Polyethylene (PE) insulated cables have been installed from the sixties in Continental Europe and the United States but failure rates were high and the use of cross-linked polyethylene (XLPE) insulation is now normal. Cables laid are either three-core XLPE as shown in Fig. 14.3 or single-core XLPE as shown in Fig. 14.4.

In Fig. 14.4 it can be seen that the screen is made up of copper wires applied over the semiconducting layer on the XLPE insulation. This has to be replaced with copper tape where fault levels are high.

As a further note about fault location, it should be remembered that an open circuit screen or sheath gives the same PE/TDR display trace feature as an open circuit core. Insulation thicknesses and stresses for polymeric cables are given in Table 14.2.

Fig. 14.2 Three-core, 150 mm², 19/33 kV sL cable (from *Electric Cables Handbook*, p. 286, ed., D. McAllister, Granada, London, 1982)

Medium voltage single-core cables can be laid in flat, side-by-side formation or bound in trefoil.

INFORMATION ON CABLES 299

Fig. 14.3 Construction of a three-core, 8.7/15 kV, XLPE insulated, steel wire armoured cable (from *Electric Cables Handbook*, p. 334, ed., D. McAllister, Granada, London, 1982)

Table 14.1 Insulation thickness (minimum) and electrical stress (maximum) for paper-insulated cables (from *Electric Cables Handbook*, p. 287, ed., D. McAllister, Granada, London, 1982)

Voltage (kV)	Conductor size (mm^2)	Belted design insulation thickness		Single-core and screened design	
		Between conductors (mm)	Conductor sheath (mm)	Insulation (mm)	Stress (MV/m)
0.6/1	50	1.4	1.2	1.2	0.24
	1000			2.0	0.13
1.9/3.3	50	2.4	1.8	1.8	1.3
	1000			2.0	1.0
3.8/6.6	50	4.2	2.7	2.4	2.0
	1000			2.4	1.7
6.35/11	50	5.6	3.4	2.8	2.9
	1000			2.8	2.4
8.7/15	50			3.6	3.3
	1000			3.6	2.6
12.7/22	50			4.9	3.8
	1000			4.9	2.8
19/33	50			7.3	4.4
	1000			6.8	3.2

300 UNDERGROUND CABLE FAULT LOCATION

Fig. 14.4 Construction of a single-core 18/30 kV XLPE insulated cable (from *Electric Cables Handbook*, p. 335, ed., D. McAllister, Granada, London, 1982)

Low voltage cables can be:

1 Four (occasionally five)-core paper-insulated, lead covered, steel tape or wire armoured with full size or reduced cross-section neutral such as that shown in Fig.14.5.
2 Four-core PVC insulated with steel wire armouring and PVC sheath overall, as shown in Fig.14.6.
3 Three-core combined neutral and earth (CNE) cable where the sheath or wire screen acts as neutral as well as earth return. These cables are only used on multiple earthed systems where the neutral is earthed at points other than at the source substation, e.g. tee, straight and stop end joints, and at the ends of runs.

In the United Kingdom almost all low voltage systems are multiple earthed and, for the most part, the only cables being laid and jointed are

Table 14.2 Insulation thicknesses and stress for polymeric cables (from *Electric Cables Handbook*, p. 333, ed., D. McAllister, Granada, London, 1982)

Rated voltage	Insulation thickness (mm)			Electrical stress (kV/mm) 185 mm² conductor	
	PE	XLPE	EPR	Maximum	Minimum
3.6/6*	2.5	2.5	3.0	1.63	1.28
6/10	3.4	3.4	3.4	2.07	1.52
8.7/15	4.5	4.5	4.5	2.38	1.60
12/20	5.5	5.5	5.5	2.79	1.74
18/30	8.0	8.0	8.0	3.12	1.67

* These figures are true for conductors up to 185 mm². Above this size the thickness increases.

INFORMATION ON CABLES 301

Fig. 14.5 Four-core, 70 mm², 600/1000 V paper-insulated lead sheathed cable with STA and bituminous finish (from *Electric Cables Handbook*, p. 284, ed., D. McAllister, Granada, London, 1982)

Fig. 14.6 Four-core, 600/1000 V copper conductor, PVC insulated SWA cable with extruded bedding (from *Electric Cables Handbook*, p. 295, ed., D. McAllister, Granada, London, 1982)

Fig. 14.7 Consac type of CNE cable (from *Electric Cables Handbook*, p. 308, ed., D. McAllister, Granada, London, 1982)

aluminium cored CNE ones. The networks, however, are totally hybrid with regard to cable type as there are still very many kilometres of four-core cables in circuit. Examples of CNE cables are shown in Figs 14.7 and 14.8.

In the United States the URD (underground residential distribution) cables used for primary distribution at voltages between 2.4 and 34.5 kV typically have single or twin aluminium cores with extruded core insulation and helically wound copper wires as a concentric neutral. Some systems have a bare copper neutral laid separately.

Early installations suffered failures both from corrosion of the neutral cores and the formation of water trees in the early **HMWPE** (high molecular

Fig. 14.8 Waveconal CNE cable with XLPE insulation (from *Electric Cables Handbook*, p. 309, ed., D. McAllister, Granada, London, 1982)

density thermoplastic polyethylene) and XLPE (cross-linked polyethylene) insulation.

Figure 14.9 shows a water tree with an electrical tree forming within it. Figure 14.10 shows early URD cable construction and Fig. 14.11 a moisture-resistant version. Both HMWPE and early XLPE are more susceptible to treeing. EPR (ethylene propylene rubber) and later tree retardant XLPE are less so.

Water-impervious URD cables are also made with a water blocking tape overlaid by a plastic-clad aluminium moisture barrier under the PE sheath. Cables with similar water-impervious construction are also used in Europe.

With regard to extruded insulation in general, XLPE has advantages over PVC (polyvinyl chloride) in that it is lighter, has superior dielectric properties and current ratings and is more moisture resistant. Other properties of XLPE are:

- Low deformation below 100°C
- Low dissipation factor
- Low dielectric constant
- Physically tough
- Chemical and oil resistant

The properties of EPR include:

- Resistance to heat with low deformation above 100°C
- Low thermal expansion
- Low shrinkback
- Flexible
- Corona resistant
- Tree resistant
- Resistant to sunlight

14.3 Voltage withstand of cables

It is not always the fault location engineer who has to carry out the routine proof testing of cables on commissioning or for maintenance purposes. Nonetheless, the engineer can encounter the problem of deciding what level

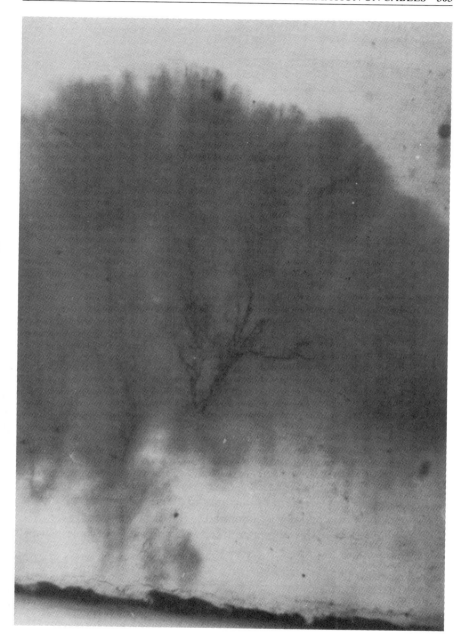

Fig. 14.9 Electrical tree forming within a water tree (from *Electric Cables Handbook*, p. 329, ed., D. McAllister, Granada, London, 1982)

304 UNDERGROUND CABLE FAULT LOCATION

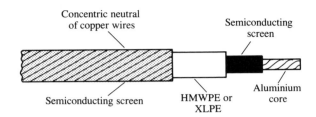

Fig. 14.10 Early URD cable construction

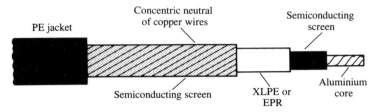

Fig. 14.11 A moisture-resistant version of URD cable construction

of voltage can be applied to a cable during fault location without causing damage.

The straightforward safeguard is to remain within the limits set down by the national or supranational test standards applicable in the country concerned. Some are more stringent than others, and some require phase-to-phase tests at the lower medium voltages as well as the phase-to-earth test. While stressing once more that the statutory limits should be adhered to, the values set down in Table 14.3 can be used as a rough guide to MV test levels.

It can be seen from this that the values are approximately four times U (nominal phase-to-earth voltage). The following points should be noted:

1 Test values must be reduced, sometimes to about half their proper value, if old or suspect switchgear is connected to the cable.
2 It is good practice only to apply a d.c. test voltage to a cable which is

Table 14.3 A guide to MV test levels

Nominal voltage (kV)	D.c. test voltage phase to earth (kV)	Duration (min)
3.8/6.6	15	15
6.35/11	25	15
8.7/15	37	15
12.7/22	50	15
19.0/33	76	15

isolated from the system voltage by a 'double break', i.e. not just one open switch.
3 It should always be borne in mind that there is a probability of voltage doubling occurring while carrying out surge and burn-down tests.

14.4 Useful tables

The following are tables containing information and values which, although obtainable from other sources, may prove useful gathered together in one volume.

Table 14.4 International statistics—annealed copper stranded conductors (from *BICC Electrical Conductors*, Table 2, pp. 32–37, British Insulated Callender's Cables Ltd, London, 1957)

Nominal area			B & SG (AWG)	Stranding and wire diameter		Approximate overall (equivalent) diameter		Calculated electrical area			Nominal weight		Standard resistance at 20°C (68°F) (plain wire)		
inch²	mm²	circ mils		inch	mm	inch	mm	inch²	mm²	circ mils	lb/1000 yd	kg/km	ohm/1000 yd	ohm/km	inch²
—	—	13 090	9	1/.1144	1/2.91	0.1144	2.91	0.010 28	6.634	13 090	118.8	58.93	2.377	2.600	—
—	—	13 090	9	7/.0432	7/1.10	0.130	3.30	0.010 09	6.508	12 840	120.7	59.86	2.422	2.649	—
0.01	—	—	—	7/.044	7/1.12	0.132	3.35	0.010 46	6.751	13 320	125.2	62.11	2.335	2.555	0.01
—	—	16 510	8	1/.1285	1/3.26	0.128	3.26	0.012 97	8.366	16 510	150.0	74.36	1.884	2.061	—
—	—	16 510	8	7/.0486	7/1.23	0.146	3.70	0.012 77	8.237	16 260	152.7	75.75	1.914	2.093	—
0.0145	—	—	—	7/.052	7/1.32	0.156	3.96	0.014 62	9.430	18 610	174.8	86.71	1.672	1.828	0.0145
—	10	—	—	1/.141	1/3.58	0.141	3.58	0.015 61	10.07	19 880	180.5	89.54	1.565	1.711	—
—	10	—	—	7/.054	7/1.37	0.162	4.12	0.015 76	10.17	20 070	188.5	93.51	1.550	1.695	—
—	—	20 820	7	1/.1443	1/3.67	0.144	3.67	0.016 35	10.55	20 820	189.1	93.80	1.494	1.634	—
—	—	20 820	7	7/.0545	7/1.38	0.164	4.15	0.016 06	10.36	20 440	192.0	95.24	1.522	1.664	—
—	—	26 240	6	1/.162	1/4.11	0.162	4.11	0.020 61	13.30	26 240	238.3	118.2	1.185	1.296	—
—	—	26 240	6	7/.0612	7/1.55	0.184	4.66	0.020 25	13.06	25 780	242.2	120.1	1.207	1.320	—
0.0225	—	—	—	7/.064	7/1.63	0.192	4.88	0.022 14	14.28	28 190	264.9	131.4	1.104	1.207	0.0225
—	16	—	—	7/.068	7/1.73	0.204	5.18	0.024 99	16.12	31 820	299.0	148.3	0.9777	1.069	—
—	—	33 090	5	7/.0688	7/1.75	0.206	5.24	0.025 59	16.51	32 580	306.0	151.8	0.9550	1.044	—
0.03	—	—	—	19/.044	19/1.12	0.220	5.59	0.028 35	18.29	36 100	340.4	168.9	0.8619	0.9425	0.03
—	—	41 740	4	7/.0772	7/1.96	0.232	5.88	0.032 22	20.78	41 020	385.3	191.1	0.7585	0.8295	—
—	25	—	—	7/.085	7/2.16	0.255	6.48	0.039 05	25.19	49 720	467.1	231.7	0.6257	0.6843	—
0.04	25	—	—	19/.052	19/1.32	0.260	6.60	0.039 60	25.55	50 420	475.3	235.8	0.6171	0.6748	0.04
—	—	52 620	3	7/.0867	7/2.20	0.260	6.61	0.040 63	26.21	51 730	486.0	241.1	0.6014	0.6577	—
—	—	66 360	2	7/.0974	7/2.47	0.292	7.42	0.051 28	33.08	65 290	613.3	304.2	0.4765	0.5211	—
—	35	—	—	7/.100	7/2.54	0.300	7.62	0.054 05	34.87	68 820	646.5	320.7	0.4521	0.4944	—

INFORMATION ON CABLES   307

Table 14.4 (cont.)

Nominal area			B & SG (AWG)	Stranding and wire diameter		Approximate overall (equivalent) diameter		Calculated electrical area		circ mils	Nominal weight		Standard resistance at 20°C (68°F) (plain wire)		
inch²	mm²	circ mils		inch	mm	inch	mm	inch²	mm²		lb/1000 yd	kg/km	ohm/1000 yd	ohm/km	inch²
—	35	—	—	19/.061	19/1.55	0.305	7.75	0.054 50	35.16	69 390	654.2	324.5	0.4484	0.4904	—
0.06	—	—	—	19/.064	19/1.63	0.320	8.13	0.059 99	38.70	76 380	720.2	357.3	0.4074	0.4455	0.06
—	—	83 690	1	19/.0664	19/1.69	0.332	8.43	0.064 57	41.66	82 210	775.1	384.5	0.3784	0.4139	—
0.075	50	—	—	19/.072	19/1.83	0.360	9.14	0.075 92	48.98	96 660	911.4	452.1	0.3219	0.3520	0.075
—	—	—	—	19/.073	19/1.85	0.365	9.27	0.078 05	50.35	99 380	936.9	464.8	0.3131	0.3424	—
—	—	105 600	1/0	19/.0745	19/1.89	0.373	9.46	0.081 29	52.44	103 500	975.8	484.1	0.3006	0.3288	—
0.1	—	—	—	19/.083	19/2.11	0.415	10.5	0.1009	65.09	128 500	1212	601.2	0.2422	0.2649	0.1
—	70	133 100	2/0	19/.0837	19/2.13	0.419	10.6	0.1026	66.19	130 600	1232	611.1	0.2382	0.2605	—
—	—	—	—	19/.086	19/2.18	0.430	10.9	0.1083	69.87	137 900	1300	645.0	0.2256	0.2467	—
0.12	—	—	—	37/.064	37/1.63	0.448	11.4	0.1168	75.33	148 700	1403	696.0	0.2093	0.2289	0.12
—	—	167 800	3/0	19/.094	19/2.39	0.470	11.9	0.1294	83.49	164 800	1553	770.4	0.1888	0.2065	—
—	—	167 800	3/0	37/.0673	37/1.71	0.470	11.9	0.1291	83.29	164 400	1551	769.4	0.1893	0.2070	—
—	95	—	—	19/.101	19/2.57	0.505	12.8	0.1494	96.39	190 200	1793	889.4	0.1636	0.1789	—
0.15	95	—	—	37/.072	37/1.83	0.504	12.8	0.1478	95.34	188 100	1776	881.0	0.1654	0.1808	0.15
—	—	211 600	4/0	19/.1055	19/2.68	0.528	13.4	0.1630	105.2	207 500	1957	970.8	0.1499	0.1639	—
—	120	—	—	37/.081	37/2.06	0.567	14.4	0.1870	120.7	238 100	2247	1115	0.1307	0.1429	—
—	—	250 000	—	37/.0822	37/2.09	0.575	14.6	0.1926	124.3	245 200	2314	1148	0.1269	0.1388	—
0.20	—	—	—	37/.083	37/2.11	0.581	14.8	0.1964	126.7	250 000	2360	1171	0.1244	0.1361	0.20
—	150	300 000	—	37/.090	37/2.29	0.630	16.0	0.2309	149.0	294 000	2774	1376	0.1058	0.1157	—
0.25	—	—	—	37/.093	37/2.36	0.651	16.5	0.2465	159.1	313 900	2963	1470	0.099 11	0.1084	0.25
—	—	350 000	—	37/.0973	37/2.47	0.681	17.3	0.2699	174.1	343 600	3243	1609	0.090 55	0.099 03	—
—	185	—	—	37/.100	37/2.54	0.700	17.8	0.2851	183.9	363 000	3426	1699	0.085 72	0.093 75	—
0.3	—	—	—	37/.103	37/2.62	0.721	18.3	0.3024	195.1	385 000	3634	1803	0.080 81	0.088 37	0.3

308 UNDERGROUND CABLE FAULT LOCATION

Table 14.4 (cont.)

Nominal area			Stranding and wire diameter			Approximate overall (equivalent) diameter			Calculated electrical area		Nominal weight				Standard resistance at 20°C (68°F) (plain wire)	
inch²	mm²	B & SG (AWG)	circ mils	inch	mm	inch	mm		inch²	mm²	circ mils	lb/ 1000 yd	kg/km		ohm/ 1000 yd	ohm/km
—	—	400 000	—	37/.104	37/2.64	0.728	18.5	0.3083	198.9	392 500	3705	1838		0.079 26	0.086 68
—	240	—	—	37/.114	37/2.90	0.798	20.3	0.3705	239.0	471 700	4452	2208		0.065 96	0.072 14
—	240	—	—	61/.089	61/2.26	0.801	20.3	0.3722	240.0	473 900	4474	2219		0.065 66	0.071 81
—	—	500 000	—	37/.1162	37/2.95	0.813	20.7	0.3849	248.3	490 100	4625	2294		0.063 49	0.069 43
—	—	500 000	—	61/.0905	61/2.30	0.814	20.7	0.3848	248.3	490 000	4626	2295		0.063 50	0.069 44
0.4	—	—	—	61/.093	61/2.36	0.837	21.3	0.4064	262.2	517 400	4885	2423		0.010 13	0.065 76
—	300	—	—	61/.099	61/2.51	0.891	22.6	0.4605	297.1	586 500	5536	2746		0.053 06	0.058 03
—	—	600 000	—	61/.0992	61/2.52	0.893	22.7	0.4624	298.3	588 700	5558	2757		0.052 85	0.057 80
0.5	—	—	—	61/.103	61/2.62	0.927	23.5	0.4985	321.6	634 700	5992	2972		0.049 02	0.053 61
—	—	700 000	—	61/.1071	61/2.72	0.964	24.5	0.5389	347.7	686 200	6479	3214		0.045 34	0.049 59
—	—	750 000	—	61/.1109	61/2.82	0.998	25.4	0.5779	372.8	735 800	6947	3446		0.042 29	0.046 25
—	—	750 000	—	91/.0908	91/2.31	0.999	25.4	0.5778	372.8	735 700	6948	3447		0.042 29	0.046 25
0.6	—	—	—	91/.093	91/2.36	1.023	26.0	0.6062	391.1	771 800	7289	3616		0.040 32	0.044 09
—	400	—	—	61/.114	61/2.90	1.026	26.1	0.6106	393.9	777 400	7341	3642		0.040 02	0.043 77
—	—	800 000	—	61/.1145	61/2.91	1.031	26.2	0.6160	397.4	784 300	7405	3673		0.039 67	0.043 38
—	—	800 000	—	91/.0938	91/2.38	1.032	26.2	0.6166	397.8	785 100	7414	3678		0.039 63	0.043 34
0.75	—	—	—	91/.103	91/2.62	1.133	28.8	0.7435	479.7	946 700	8940	4435		0.032 87	0.035 94
—	500	1 000 000	—	61/.1280	61/3.25	1.152	29.3	0.7698	496.6	980 100	9254	4590		0.031 74	0.034 72
—	—	1 000 000	—	91/.1048	91/2.66	1.153	29.3	0.7697	496.6	980 100	9255	4591		0.031 75	0.034 72
0.85	—	—	—	127/.093	127/2.36	1.209	30.7	0.8459	545.8	1 077 000	10 173	5046		0.028 89	0.031 59
—	625	—	—	91/.117	91/2.97	1.287	32.7	0.9594	619.0	1 222 000	11 536	5722		0.025 47	0.027 86

INFORMATION ON CABLES

Table 14.4 (cont.)

Nominal area			Stranding and wire diameter		Approximate overall (equivalent) diameter		Calculated electrical area		Nominal weight			Standard resistance at 20°C (68°F) (plain wire)	
$inch^2$	mm^2	circ mils	B & SG (AWG)	inch	mm	inch	mm	$inch^2$	mm^2	circ mils	lb/ 1000 yd	kg/km	ohm/ 1000 yd	ohm/km $inch^2$
—	—	1 250 000	—	91/.1172	91/2.98	1.289	32.7	0.9627	621.1	1 226 000	11 575	5742	0.025 38	0.027 76 —
—	—	1 250 000	—	127/.0992	127/2.52	1.290	32.8	0.9625	620.9	1 225 000	11 574	5741	0.025 39	0.027 77 —
1.0	—	1 500 000	—	127/.103	127/2.62	1.339	34.0	1.0376	669.4	1 321 000	12 478	6190	0.023 55	0.025 75 1.0
—	—	1 500 000	—	91/.1284	91/3.26	1.412	35.9	1.155	745.5	1 471 000	13 893	6892	0.021 15	0.023 13 —
—	—	1 500 000	—	127/.1087	127/2.76	1.413	35.9	1.156	745.6	1 471 000	13 897	6894	0.021 15	0.023 12 —
—	800	—	—	91/.132	91/3.35	1.452	36.9	1.221	787.7	1 555 000	14 683	7284	0.020 01	0.021 88 —
1.25	—	—	—	127/.112	127/2.84	1.456	37.0	1.227	791.5	1 562 000	14 754	7319	0.019 92	0.021 78 1.25
1.5	—	—	—	169/.107	169/2.72	1.605	40.8	1.490	961.3	1 897 000	17 920	8889	0.016 40	0.017 94 1.5
—	1000	—	—	91/.147	91/3.73	1.617	41.1	1.514	976.8	1 928 000	18 210	9033	0.016 14	0.017 65 —
—	—	2 000 000	—	127/.1255	127/3.19	1.632	41.5	1.540	993.8	1 961 000	18 525	9189	0.015 86	0.017 35 —
—	—	2 000 000	—	169/.1088	169/2.76	1.632	41.5	1.541	993.9	1 962 000	18 528	9191	0.015 86	0.017 35 —

BRITISH SIZES
  BS 7: 1953 nd BS 480: 1954
B & SG (AWG) SIZES
  Based on CESA C68A; ASTM B8–53 and IPCEA 5–19–8 Classes B
  & C (preferred sizes)
METRIC SIZES
  VDE 0255/51 and 0265/52

BRITISH SIZES
  The following tolerances on resistance are permitted in British Standard Specifications:

	per cent
Single wires, tinned, below .036-inch diameter	+5
Single wires, tinned, .036-inch diameter and above	+4
Single wires, plain	+3
Stranded conductor, tinned, below .036-inch diameter	+4
Stranded conductor, tinned, .036-inch diameter and above	+3
Stranded conductor, plain	+2

A further increase in resistance of 2 per cent is allowable for the laying-up of twin and multicore cables

Table 14.5 International statistics—hard-drawn copper stranded conductors (from *BICC Electrical Conductors*, Table 3, pp. 38–41, British Insulated Callender's Cables Ltd, London, 1957)

Nominal area			Stranding and wire diameter			Approximate overall diameter		Actual area			Calculated electrical area			Nominal weight		Standard resistance at 20 °C (68 °F)		Nominal breaking load		Nominal area	
inch²	mm²	circ mils	B & SG (AWG)	inch	mm	inch	mm	inch²	mm²	circ mils	inch²	mm²	circ mils	lb/1000 yd	kg/km	ohm/1000 yd	ohm/km	lb	kg	inch²	mm²
0.22	10	—	—	7.0532	7/1.35	0.159	4.05	0.01556	10.04	19810	0.01543	9.956	19650	181.4	89.98	1.630	1.783	950	430	0.22	—
	16	—	—	7.064	7/1.63	0.192	4.88	0.02252	14.53	28670	0.02233	14.41	28430	262.5	130.2	1.126	1.232	1370	620		
		—	—	7.067	7/1.7	0.201	5.1	0.02461	15.88	31330	0.02440	15.74	31070	286.9	142.4	1.031	1.127	1490	675		
0.25	—	—	—	3.104	3/2.64	0.224	5.67	0.02548	16.44	32450	0.02529	16.32	32200	297.0	147.3	0.9942	1.088	1520	690	0.25	—
	25	41740	4	3.1180	3/3.0	0.254	6.45	0.03286	21.20	41840	0.03261	21.04	41520	383.0	190.0	0.7709	0.8431	1930	875		
		—	—	7.7087	7/2.1	0.248	6.3	0.03760	24.26	47880	0.03729	24.06	47480	438.4	217.4	0.6742	0.7373	2260	1025		
		52620	3	3.1325	3/3.37	0.285	7.24	0.04137	26.69	52670	0.04105	26.48	52270	482.1	239.1	0.6122	0.6695	2400	1090		
0.5	—	—	—	3.147	3/3.73	0.317	8.05	0.05092	32.85	64830	0.05052	32.60	64320	593.4	294.4	0.4973	0.5438	2920	1320	0.5	—
		—	—	3.1487	3/3.78	0.320	8.13	0.05210	33.61	66340	0.05170	33.35	65830	607.2	301.2	0.4859	0.5314	2980	1350		
	35	—	—	7.0984	7/2.5	0.295	7.50	0.05323	34.34	67780	0.05279	34.06	67210	620.6	307.8	0.4761	0.5207	3180	1440		
0.58	—	—	—	7.104	7/2.64	0.312	7.92	0.05946	38.36	75710	0.05897	38.05	75080	693.2	343.9	0.4261	0.4661	3540	1605	0.58	—
		83690	1	3.167	3/4.24	0.360	9.14	0.06571	42.39	83670	0.06521	42.07	83030	765.8	379.9	0.3851	0.4211	3700	1675		
0.75		—	—	7.116	7/2.95	0.348	8.84	0.07398	47.73	94190	0.07335	47.33	93420	862.4	427.8	0.3424	0.3745	4350	1975	0.75	—
	50	—	—	7.1181	7/3.0	0.354	9.0	0.07668	49.47	97630	0.07605	49.06	96830	893.9	443.4	0.3304	0.3613	4510	2045		
	50	—	—	19.0709	19/1.8	0.354	9.0	0.07501	48.40	95510	0.07411	47.81	94360	878.3	435.7	0.3395	0.3713	4440	2015		
		105600	1/0	7.1228	7/3.12	0.368	9.35	0.08291	53.49	105560	0.08222	53.04	104700	966.5	479.4	0.3056	0.3342	4860	2205		
0.10	—	—	—	7.136	7/3.45	0.408	10.4	0.1017	65.60	129470	0.1008	65.06	128300	1186	588.1	0.2490	0.2723	5870	2660	0.10	—
		—	—	19.0827	19/2.1	0.413	10.5	0.1021	65.85	129950	0.1008	65.05	128300	1195	592.8	0.2495	0.2728	6010	2725		
	70	—	—	7.1379	7/3.5	0.414	10.5	0.1044	67.35	132920	0.1035	66.79	131800	1217	603.7	0.2425	0.2652	6010	2725		
		133100	2/0	7.1548	7/3.93	0.464	11.8	0.1317	85.00	167740	0.1307	84.29	166400	1536	761.9	0.1921	0.2101	7510	3405		
	95	167800	3/0	19.0984	19/2.5	0.492	12.5	0.1445	93.22	183970	0.1427	92.09	181700	1692	839.2	0.1762	0.1927	8450	3830		
0.15	—	—	—	7.166	7/4.22	0.498	12.6	0.1515	97.74	192890	0.1502	96.93	191200	1766	876.1	0.1670	0.1827	8530	3870	0.15	—
		211600	4/0	7.1739	7/4.42	0.522	13.3	0.1663	107.3	211690	0.1649	106.4	210000	1938	961.5	0.1522	0.1664	9320	4225		
	120	—	—	19.1102	19/2.8	0.551	14.0	0.1812	116.9	230740	0.1790	115.5	227900	2122	1053	0.1403	0.1534	10500	4760		
		250000	—	12.1443	12/3.66	0.600	15.2	0.1963	126.6	249870	0.1939	125.1	246900	2290	1136	0.1321	0.1445	11020	4995		
0.20	—	—	—	7.193	7/4.9	0.579	14.7	0.2048	132.1	260740	0.2031	131.0	258600	2387	1184	0.1235	0.1351	11300	5110	0.20	—
0.20		—	—	19.1116	19/2.95	0.580	14.7	0.2008	129.5	255660	0.1984	128.0	252600	2351	1166	0.1267	0.1386	11600	5250	0.20	—
	150	—	—	37.0886	37/2.25	0.620	15.8	0.2281	147.2	290450	0.2250	145.1	286500	2673	1326	0.1118	0.1222	13400	6080		
		300000	—	12.1581	12/4.02	0.657	16.7	0.2356	152.0	299950	0.2327	150.1	296300	2749	1364	0.1100	0.1203	13060	5920		
0.25	—	—	—	7.215	7/5.46	0.645	16.4	0.2541	163.9	323580	0.2520	162.6	320900	2963	1470	0.09948	0.1088	13800	6250	0.25	—
0.25		—	—	19.131	19/3.33	0.655	16.6	0.2561	165.2	326060	0.2530	163.2	322100	2998	1487	0.09931	0.1086	14600	6600	0.25	—
		350000	—	12.1708	12/4.34	0.710	18.0	0.2749	177.4	350070	0.2716	175.2	345800	3243	1608	0.09420	0.1030	15100	6850		

INFORMATION ON CABLES 311

Table 14.5 (cont.)

Nominal area			Stranding and wire diameter		Approximate overall diameter		Actual area		Calculated electrical area			Nominal weight		Standard resistance at 20°C (68°F)			Nominal breaking load		Nominal area
mm²	circ mils	B & SG (AWG)	inch	mm	inch	mm	inch²	mm²	circ mils	inch²	mm²	lb/1000 yd	kg/km	ohm/1000 yd	ohm/km		lb	kg	inch²
185	—	—	37/.0984	37/2.5	0.689	17.5	0.2814	181.5	358 260	0.2775	179.0	3298	1636	0.090 59	0.099 07		16 450	7460	—
—	—	—																	0.3*
—	400 000	—	19/.144	19/3.66	0.720	18.3	0.3094	199.6	393 980	0.3057	197.2	3623	1797	0.082 17	0.089 86		17 370	7880	—
—	—	—	19/.1451	19/3.69	0.726	18.4	0.3142	202.7	400 000	0.3104	200.2	3678	1825	0.080 93	0.088 51		17 640	8000	—
240	—	—	61/.0886	61/2.25	0.797	20.2	0.3761	242.6	478 850	0.3688	237.9	4434	2199	0.068 20	0.074 58		22 100	10 020	—
—	500 000	—	19/.1622	19/4.12	0.811	20.6	0.3926	253.3	499 870	0.3878	250.2	4597	2280	0.064 74	0.070 80		21 700	9855	—
0.4*	—	—																	0.4*
—	—	—	19/.166	19/4.22	0.830	21.1	0.4112	265.3	523 560	0.4062	262.1	4814	2388	0.061 80	0.067 59		22 600	10 260	—
300	—	—	61/.984	61/2.5	0.886	22.5	0.4639	299.3	590 640	0.4549	293.5	5469	2713	0.055 28	0.060 46		27 100	12 300	—
—	600 000	—	37/.1273	37/3.23	0.891	22.6	0.4709	303.8	599 690	0.4644	299.6	5518	2737	0.054 09	0.059 16		26 900	12 200	—
0.5*	—	—																	0.5*
—	—	—	19/.185	19/4.70	0.925	23.5	0.5107	329.5	650 280	0.5045	325.5	5980	2966	0.049 74	0.054 40		27 700	12 560	—
—	700 000	—	37/.1375	37/3.49	0.962	24.4	0.5494	354.5	699 530	0.5418	349.6	6438	3194	0.046 35	0.050 69		30 900	14 040	—
—	750 000	—	37/.1424	37/3.62	0.997	25.3	0.5893	380.2	750 280	0.5811	374.9	6905	3425	0.043 21	0.047 26		33 100	15 040	—
0.6*	—	—																	0.6*
—	—	—	37/.144	37/3.66	1.008	25.6	0.6026	388.8	767 230	0.5943	383.4	7061	3503	0.042 25	0.046 21		33 800	15 350	—
—	800 000	—	37/.147	37/3.73	1.029	26.1	0.6280	405.1	799 530	0.6193	399.5	7359	3650	0.040 55	0.044 34		35 200	15 960	—
0.75*	—	—	37/.162	37/4.11	1.134	28.8	0.7626	492.0	971 030	0.7521	485.2	8937	4433	0.033 37	0.036 50		42 200	19 150	0.75*
—	1 000 000	—	37/.1644	37/4.18	1.151	29.2	0.7854	506.7	1 000 000	0.7746	499.7	9204	4566	0.032 40	0.035 44		43 300	19 640	—

* Additional to BS 125: 1954

BRITISH SIZES —— BS 125: 1954
B & SG (AWG) SIZES —— ASTM B8-53. Preferred sizes only (class AA)
METRIC SIZES —— DIN 48201: 1942

Resistance, weight and tensile strength of all sizes are calculated from the strand diameter in accordance with BS 125: 1954

312 UNDERGROUND CABLE FAULT LOCATION

Table 14.6 US copper and aluminium stranded conductor resistances at 20 °C (from *Electric Cables Handbook*, p. 627–629, ed., D. McAllister, Granada, London, 1982)

Nominal area (AWG or MCM*)	Equivalent metric area† (mm²)	Nominal d.c. resistance			Maximum d.c. resistance‡ Single-core cable		Maximum d.c. resistance‡ Multicore cable	
		Copper (ohm/km)	Coated copper (ohm/km)	Aluminium (ohm/km)	Copper (ohm/km)	Aluminium (ohm/km)	Copper (ohm/km)	Aluminium (ohm/km)
20	0.519	33.9	36.0		34.6		35.3	
18	0.823	21.4	22.7		21.8		22.2	
16	1.31	13.4	14.3		13.7		14.0	
14	2.08	8.45	8.78		8.62		8.79	
13	2.63	6.69	6.96		6.82		6.96	
12	3.31	5.32	5.53	8.71	5.43	8.88	5.54	9.06
11	4.17	4.22	4.39	6.92	4.30	7.06	4.39	7.20
10	5.26	3.34	3.48	5.48	3.41	5.59	3.48	5.70
9	6.63	2.65	2.76	4.35	2.70	4.44	2.75	4.53
8	8.37	2.10	2.19	3.45	2.14	3.52	2.18	3.59
7	10.6	1.67	1.73	2.73	1.70	2.78	1.73	2.84
6	13.3	1.32	1.38	2.17	1.35	2.21	1.38	2.25
5	16.8	1.05	1.09	1.72	1.07	1.75	1.09	1.79
4	21.2	0.832	0.865	1.36	0.849	1.39	0.866	1.42
3	26.7	0.660	0.686	1.08	0.673	1.10	0.686	1.12
2	33.6	0.523	0.544	0.857	0.533	0.874	0.544	0.891
1	42.4	0.415	0.431	0.680	0.423	0.694	0.431	0.708
1/0	53.5	0.329	0.342	0.539	0.336	0.550	0.343	0.561
2/0	67.4	0.261	0.271	0.428	0.266	0.437	0.271	0.446
3/0	85.0	0.207	0.215	0.339	0.211	0.336	0.215	0.343
4/0	107	0.164	0.169	0.269	0.167	0.274	0.170	0.279
250	127	0.139	0.144	0.228	0.142	0.233	0.149	0.238
300	152	0.116	0.120	0.190	0.118	0.193	0.120	0.197
350	177	0.0992	0.103	0.163	0.101	0.166	0.103	0.169
400	203	0.0868	0.893	0.142	0.0885	0.145	0.0903	0.148
450	228	0.0771	0.0794	0.126	0.0786	0.129	0.0802	0.132
500	253	0.0694	0.0714	0.114	0.0708	0.116	0.0722	0.118
550	279	0.0631	0.0656	0.103	0.0644	0.105	0.0657	0.107
600	304	0.0578	0.0602	0.0948	0.0590	0.0967	0.0602	0.0986
650	329	0.0534	0.0550	0.0875	0.0545	0.0893	0.0556	0.0911
700	355	0.0496	0.0510	0.0813	0.0506	0.0829	0.0516	0.0846
750	380	0.0463	0.0476	0.0759	0.0472	0.0774	0.0481	0.0789
800	405	0.0434	0.0447	0.0711	0.0443	0.0725	0.0452	0.0740
900	456	0.0386	0.0397	0.0632	0.0394	0.0645	0.0402	0.0658
1000	507	0.0347	0.0357	0.0569	0.0354	0.0580	0.0361	0.0592
1100	557	0.0316	0.0325	0.0517	0.0322	0.0527	0.0328	0.0538
1200	608	0.0289	0.0298	0.0474	0.0295	0.0483	0.0301	0.0493
1250	633	0.0278	0.0286	0.0455	0.0284	0.0464	0.0290	0.0473
1300	659	0.0267	0.0275	0.0438	0.0272	0.0447	0.0277	0.0456
1400	709	0.0248	0.0255	0.0406	0.0253	0.0414	0.0258	0.0422

The data is based on IPCEA 5–66–524, NEMA WC7 and IPCEA 5–19–81, NEMA WC3 for concentric stranded and compact stranded conductors. Different values apply to solid conductors.
* AWG up to 4/0, MCM from 250 upwards.
† Based on nominal area and incorrect for equivalent resistance.
‡ Taken as 2% above nominal resistance in case of single-core cable and 2% above single-core cable for multicore cable.

Table 14.7 Metric conductor resistances at 20 °C (from *Electric Cables Handbook*, p. 625, ed., D. McAllister, Granada, London, 1982)

Conductor size (mm^2)	Maximum d.c. resistance			Conductor size (mm^2)	Maximum d.c. resistance		
	Plain copper (ohm/km)	Metal coated copper (ohm/km)	Aluminium* (ohm/km)		Plain copper (ohm/km)	Metal coated copper (ohm/km)	Aluminium* (ohm/km)
0.5	36.0	36.7		500	0.0366	0.0369	0.0605
0.75	24.5	24.8		630	0.0283	0.0286	0.0469
1	18.1	18.2		800	0.0221	0.0224	0.0367
1.5	12.1	12.2		1000	0.0176	0.0177	0.0291
2.5	7.41	7.56		1200	0.0151	0.0151	0.0247
4	4.61	4.70	7.41	1400†	0.0129	0.0129	0.0212
6	3.08	3.11	4.61	1600†	0.0113	0.0113	0.0186
10	1.83	1.84	3.08	1800‡	0.0101	0.0101	0.0165
16	1.15	1.16	1.91	2000‡	0.0090	0.0090	0.0149
25	0.727	0.734	1.20				
				1150§	0.0156		0.0258
35	0.524	0.529	0.868	1300§	0.0138		0.0228
50	0.387	0.391	0.641				
70	0.268	0.270	0.443	380§			0.0800
95	0.193	0.195	0.320	480§			0.0633
120	0.153	0.154	0.253	600§			0.0515
				740§			0.0410
150	0.124	0.126	0.206	960§			0.0313
185	0.0991	0.100	0.164	1200§			0.0250
240	0.0754	0.0762	0.125				
300	0.0601	0.0607	0.100				
400	0.0470	0.0475	0.0778				

* Includes metal-coated and metal-clad.
† Non-preferred sizes in IEC 228.
‡ Sizes used for OF cables (not in IEC 228).
§ Solid sectoral conductors (not in IEC 228 but standard for British practice, BS 6791).

Except where stated, the data are in accordance with IEC 228 and British Standards.

314 UNDERGROUND CABLE FAULT LOCATION

Table 14.8 Comparison of AWG sizes and metric areas

AWG	mm^2	AWG	mm^2
44	0.002 027	10	5.261
40	0.004 869	9	6.634
36	0.012 67	8	8.366
32	0.032 43	7	10.55
30	0.050 67	6	13.30
28	0.080 64	5	16.77
27	0.102 17	4	21.15
26	0.1281	2	33.63
24	0.2047	1	42.41
22	0.3243	4/0	107.2
20	0.5189	225 MCM	114.0
19	0.6516	250 MCM	126.7
18	0.8258	500 MCM	253.3
16	1.308	525 MCM	266.0
15	1.651	1 M MCM	506.7
14	2.082	1.1 M MCM	557.4
12	3.308	2 M MCM	1013.4

Table 14.9 Examples of characteristic impedances

(a) 0.6/1.0 kV, belted cables

Size (mm^2)	Type	Characteristic impedance (ohms)
25	Four-core stranded aluminium	18.2
16	Four-core stranded copper	19.1
35	Four-core stranded aluminium	16.7
25	Four-core stranded copper	18.2
95	Four-core stranded aluminium	13.1
70	Four-core stranded copper	14.0
185	Four-core stranded aluminium	12.0
120	Four-core stranded copper	12.5

(b) 11 kV, belted cables

Size (mm^2)	Type	Characteristic impedance (ohms)
95	Three-core stranded aluminium	24.8
70	Three-core stranded copper	27.1
185	Three-core stranded aluminium	21.0
120	Three-core stranded copper	23.3
300	Three-core stranded aluminium	18.5
185	Three-core stranded copper	21.0

INFORMATION ON CABLES 315

Table 14.10 Conductor versus armour resistances, 0.6/1.0 kV, copper conductors

Size (mm^2)	Conductor maximum resistance at 20 °C (ohms/km)	Steel wire armour, maximum resistance at 20 °C (ohms/km)	
		Two-core	Four-core
16	1.15	3.8	3.2
25	0.727	3.7	2.3
35	0.524	2.5	2.0
50	0.387	2.3	1.8
70	0.268	2.0	1.2
95	0.193	1.4	1.1
120	0.153	1.3	0.76
150	0.124	1.2	0.68
185	0.0991	0.82	0.61
240	0.0754	0.73	0.54

Table 14.11 Conductor versus armour resistances, 0.6/1.0 kV, aluminium conductors

Size (mm^2)	Conductor maximum resistance at 20 °C (ohms/km)	Steel wire armour, maximum resistance at 20 °C (ohms/km)	
		Two-core	Four-core
16	1.91	3.6	3.0
25	1.20	3.0	1.9
35	0.868	2.0	1.7
50	0.641	2.3	1.7
70	0.443	2.1	1.2
95	0.320	1.4	1.1
120	0.253	1.3	0.75
150	0.206	—	0.69
185	0.164	—	0.62
240	0.125	—	0.55

Table 14.12 Propagation velocities—power cables

Type	Velocity of propagation ($m/\mu s$)
PE and XLPE	190–200
Oil-impregnated paper with no semi-conducting layer	160–170
Oil-impregnated paper with semi-conducting layer	150–160
PVC	105–135

See Table 3.3 for the list of propagation velocities relating to telecommunications cables.

Table 14.13 US power cables, propagation factors. It is clear that values vary considerably as they also depend on the cable construction (sometimes even on the core insulation colouring in telecommunications cables!). This is the reason that, in describing methods and procedures, great emphasis has been placed on using *known correct values* or on establishing the correct value empirically. It is also to be noted that values often seem to vary in one telecommunications cable. Although this can be due to the core colour, as mentioned above, it is much more likely to derive from the differences in length between central and peripheral cores and pairs.

Insulation type	mils	kV	Gauge	PF
XLPE	345	35	1/0	0.57
XLPE		35	750 MCM	0.51
PILC		35	750 MCM	0.52
XLPE		25	1/0	0.56
XLPE	260	25	1/0	0.51
XLPE		25	#1 Cu	0.49
PILC		25	4/0	0.54
XLPE	175	15	1/0 Al	0.55
XLPE	175	15	1/0	0.51
XLPE		15	2/0	0.49
XLPE		15	4/0	0.49
XLPE		15	#1 Cu	0.56
XLPE		15	#2 Cu and Al	0.52
XLPE	260	15	#2 Al	0.53
XLPE		15	#2 Al	0.48
XLPE		15	#4 Cu	0.52
XLPE		15	500 MCM	0.53
XLPE		15	750 MCM	0.56
XLPE	260	15	750 MCM and AL	0.53
EPR	220	15	1/0	0.52
EPR	220	15	4/0	0.58
EPR		15	#2 Al	0.55
PILC		15	4/0	0.49
EPR		5	#2	0.45
EPR		5	#6	0.57
XLPE		0.6	1/0	0.62
XLPE		0.6	4/0	0.62
XLPE		0.6	#2	0.61
XLPE		0.6	#8	0.61
XLPE		0.6	#12.6 PR	0.62

INFORMATION ON CABLES 317

Table 14.14 Resistance/temperature conversion factors—constants and reciprocals of constants (for converting resistances between various temperatures and the standard temperature of 20 °C) (from *BICC Electrical Conductors*, Table E, p. 8, British Insulated Callender's Cables Ltd, London, 1957)

Temperature (°C)	Annealed high conductivity copper		Hard-drawn high conductivity copper		Hard-drawn cadmium–copper		Hard-drawn aluminium	
	Multiplier constant	Reciprocal of constant	Multiplier constant	Reciprocal of constant	Multiplier constant	Reciprocal of constant	Multiplier constant	Reciprocal of constant
5	1.0626	0.9411	1.0606	0.9429	1.0488	0.9535	1.0643	0.9396
6	1.0582	0.9450	1.0563	0.9467	1.0454	0.9566	1.0598	0.9436
7	1.0538	0.9489	1.0521	0.9505	1.0420	0.9597	1.0553	0.9476
8	1.0495	0.9528	1.0479	0.9543	1.0386	0.9628	1.0508	0.9516
9	1.0452	0.9568	1.0437	0.9581	1.0353	0.9659	1.0464	0.9557
10	1.0409	0.9607	1.0396	0.9619	1.0320	0.9690	1.0420	0.9597
11	1.0367	0.9646	1.0355	0.9657	1.0287	0.9721	1.0376	0.9637
12	1.0325	0.9686	1.0314	0.9695	1.0254	0.9752	1.0333	0.9678
13	1.0283	0.9725	1.0274	0.9733	1.0222	0.9783	1.0290	0.9718
14	1.0241	0.9764	1.0234	0.9771	1.0189	0.9814	1.0248	0.9758
15	1.0200	0.9804	1.0194	0.9810	1.0157	0.9845	1.0206	0.9798
16	1.0160	0.9843	1.0155	0.9848	1.0125	0.9876	1.0164	0.9839
17	1.0119	0.9882	1.0116	0.9886	1.0094	0.9907	1.0122	0.9879
18	1.0079	0.9921	1.0077	0.9924	1.0062	0.9938	1.0081	0.9919
19	1.0039	0.9961	1.0038	0.9962	1.0031	0.9969	1.0040	0.9960
20	1.0000	1.0000	1.0000	1.0000	1.0000	1.0000	1.0000	1.0000
21	0.9961	1.0039	0.9962	1.0038	0.9969	1.0031	0.9960	1.0040
22	0.9922	1.0079	0.9924	1.0076	0.9938	1.0062	0.9920	1.0081
23	0.9883	1.0118	0.9887	1.0114	0.9908	1.0093	0.9881	1.0121
24	0.9845	1.0157	0.9850	1.0152	0.9877	1.0124	0.9841	1.0161
25	0.9807	1.0197	0.9813	1.0191	0.9847	1.0155	0.9802	1.0202
26	0.9770	1.0236	0.9777	1.0229	0.9817	1.0186	0.9764	1.0242
27	0.9732	1.0275	0.9740	1.0267	0.9788	1.0217	0.9726	1.0282
28	0.9695	1.0314	0.9704	1.0305	0.9758	1.0248	0.9688	1.0322
29	0.9658	1.0354	0.9668	1.0343	0.9728	1.0279	0.9650	1.0363
30	0.9622	1.0393	0.9633	1.0381	0.9699	1.0310	0.9613	1.0403
35	0.9443	1.0589	0.9459	1.0572	0.9556	1.0465	0.9430	1.0605
40	0.9271	1.0786	0.9292	1.0762	0.9416	1.0620	0.9254	1.0806
45	0.9105	1.0982	0.9130	1.0953	0.9281	1.0775	0.9085	1.1008
50	0.8945	1.1179	0.8974	1.1143	0.9149	1.0930	0.8921	1.1209
55	0.8791	1.1375	0.8823	1.1334	0.9021	1.1085	0.8764	1.1411
60	0.8642	1.1572	0.8678	1.1524	0.8897	1.1240	0.8612	1.1612
65	0.8498	1.1768	0.8536	1.1715	0.8776	1.1395	0.8465	1.1814
70	0.8358	1.1965	0.8400	1.1905	0.8658	1.1550	0.8323	1.2015

The above constants are based on the standard expressions
$R_t = R_{20}[1 + \alpha(t-20)]$
and $R_{20} = \dfrac{R_t}{[1 + \alpha(t-20)]}$

where R_t = resistance at temperature $t\,°C$
R_{20} = resistance at 20 °C
α = standard resistance temperature coefficient at 20 °C

The multiplier constant in the table is calculated from $\dfrac{1}{[1 + \alpha(t-20)]}$

Table 14.15 Conversion factors and multiple metric units (from *Electric Cables Handbook*, p. 624, ed., D. McAllister, Granada, London, 1982)

Conversion factors

From	To	Factor	Reciprocal
mm	in	0.0394	25.4
m	ft	3.2808	0.3048
m	yard	1.0936	0.9144
km	mile	0.6214	1.6093
mm^2	in^2	0.001 55	645.16
mm^2	circular mil	1973.5	5.0671×10^{-4}
N	lbf	0.2248	4.4482
N	kgf	0.1020	9.8067
bar	N/m^2	10^5	10^{-5}
bar	lbf/in^2	14.5	68.9476×10^{-3}
N/m^2	torr (mmHg)	7.501×10^{-3}	133.32
N/m^2	lbf/in^2	1.450×10^{-4}	6894.76
kgf/cm^2	lbf/in^2	14.223	0.070 31
kg	ton	9.8421×10^{-4}	1016.05
kg	lb	2.2046	0.4536
t (tonne)	lb	2204.6	0.4536×10^{-3}
1 (litre)	gal (UK)	0.2202	4.541
1 (litre)	gal (US)	0.2642	3.785

Multiple and submultiple metric units

Multiple	Prefix	Symbol	Submultiple	Prefix	Symbol
10^{12}	tera	T	10^{-1}	deci	d
10^9	giga	G	10^{-2}	centi	c
10^6	mega	M	10^{-3}	milli	m
10^3	kilo	k	10^{-6}	micro	μ
10^2	hecto	h	10^{-9}	nano	n
10	deca	da	10^{-12}	pico	p

PART FIVE

Choice of equipment

In Chapters 15 and 16, where comments on equipment and costs refer to telecommunications fault location, this is made clear in the text. Otherwise, it can be assumed that all references are to power cable fault location.

15

Factors

In underground cable fault location there are many factors that influence choice of equipment, the most important not always being *cost*! Some of these are:

- Type of company/authority
- Fault incidence
- Length and cost of outages
- Type of network
- Environment
- Budget

For example, a supply authority or company has a huge network in terms of network-kilometres/miles and therefore a high fault incidence. This justifies the purchase of full sets of up to date equipment for daily use.

At the other end of the scale, even a large industrial installation will only have a small network in comparison and therefore needs much less equipment for occasional use, perhaps only every few months. It may even be cheaper and more effective to possess no equipment at all and call in consultants or contractors as necessary. Paradoxically, if such an industrial network is feeding processes and production which, when halted, means huge losses by the hour, large expenditure on fault location equipment is fully justified.

The type of network is a major factor. There may be a preponderance of a certain type of cable which dictates the type of equipment needed, for instance, for unshielded telecommunications cable on which most faults are ground contact faults.

A history of bad jointing, a policy error resulting in the installation of poor quality joints or cable, or a spate of corroding connectors in a telecommunications loop can all produce increased fault incidence in certain areas or time windows. They can also result in a particular *type* of fault predominating which, in turn, influences the type and number of instruments purchased.

With regard to environment, difficult terrain and climate have a direct influence on the choice of portable equipment versus tailor-made test vehicles with overland capabilities and controlled test cabin conditions. A serious limitation is also encountered when cables pass through hazardous areas where it is not permitted to use high voltages or currents.

More often than not, however, the bottom line is *money*. Too often the maintenance engineer tends to do battle with the manager and/or financial controller over the purchase of test equipment. Management needs to husband resources and keep expenditure down and may veto a particular purchase or insist on buying a cheaper instrument which may not be as effective as the one requested by the maintenance engineer. Also, management may think that test instruments have a very long life.

In actual fact, the maintenance engineer works for the same company as the manager and, by a judicious choice of test instruments, can save money by reducing outage times to a minimum. It is not always clear to management, however, that money can be saved because, even with little equipment, faults are always found. There can, however, be considerable waste of labour and materials if cables are cut and jointed too often during fault location.

It therefore behoves the maintenance engineer to keep records and carefully to prepare submissions for the purchase of equipment which clearly demonstrate that time and money can be saved by the purchase of the right equipment. The *capital* budget in any organization under which test equipment is purchased is always under close scrutiny but it is rarely pointed out that very significant sums of money are often wasted as *revenue* expenditure, the budget for which is always difficult to set and monitor.

Instruments have a finite life. Some wear out, some are broken or damaged but, increasingly, a new factor is entering the equation – obsolescence. These days technology is developing and changing so rapidly that instruments become out of date much more quickly than they used to. Nevertheless, neither the manager nor the financial controller will be against the purchase of equipment if they are presented with a proper case.

Furthermore, while it is normal to depreciate a company vehicle over, say, five years, this very normal accounting procedure is not always applied to the purchase and replacement of test instruments. They are, however, a capital asset and can be allocated a finite life and depreciated over that period. The financial controller will then be quite happy to provide funds for the replacement of equipment.

Regarding the actual cost of equipment, the graph in Fig. 15.1 emphasize

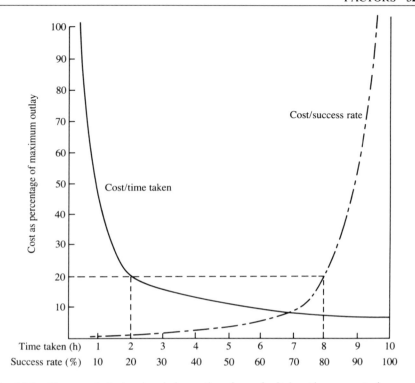

Fig. 15.1 Curves plotted using information from fault location experts in many countries in research carried out by Seba Dynatronic GmbH (reproduced by permission of Seba Dynatronic GmbH, Germany)

some simple, perhaps obvious, but very interesting facts. The curves themselves have been plotted using information supplied by many fault location experts from many countries. They are empirical but soundly based on practical knowledge. Times are calculated from the time the cable is made available to the operator to the time the location is marked on the ground. All values indicated should be treated as approximate.

Cost is given as a percentage of 'maximum outlay' because equipment costs rise over the years with inflation. 'Maximum outlay' should be taken as the cost, at a given time, of a complete, built-to-order, test vehicle together with all normal portable equipment.

From the cost/time curve it is clear that no amount of money can buy a fault location time of zero hours but, for a 50 per cent outlay, fault location times can be brought under one hour. Conversely, expenditure significantly less than 20 per cent results in long location times.

The cost/success rate curve illustrates the fact that every fault location

engineer knows – even with the very best equipment, it is impossible to achieve a 100 per cent success rate. There will always be the occasional fault that defies location.

An interesting compromise is indicated by the coordinates marked on the graph; i.e. for an outlay of 20 per cent, 80 per cent of faults can be found within two hours. This expenditure corresponds quite well with the cost of the *minimum kit* (see Chapter 16, 'choice').

It could also be said that, if 50 per cent purchases a near 90 per cent success rate and locations under one hour, why should 100 per cent ever be spent? Indeed, the purchase of a fully equipped test vehicle is only justified if fault location times are very critical (for instance where there are many radial feeders) or for other reasons, such as providing central equipment for a 'sun and planets' system (see Chapter 16, 'choice').

16

Choice

There is always a difference between the cost of equipment necessary for fast efficient fault location and the available budget. This, by definition, means that compromises have to be made.

Decisions regarding compromise should be influenced mainly by network and staff considerations. For instance, there is no point in spending 10 per cent on a top of the range PE set/TDR and sacrificing pin-pointing apparatus for fault location of an industrial network with relatively short cables. The best approach here would be to purchase a suitable surge generator for somewhat less than 10 per cent, together with basic diagnostic, pre-location and tracing equipment. Similarly, a refined instrument is no good in the hands of an operator not trained to use it.

Another aspect very much concerned with staff and training *vis-à-vis* equipment is *safety*. High voltages can be extremely dangerous, particularly those from a surge generator, when they are backed by large banks of capacitors.

16.1 Minimum kit

Minimum kit should cover diagnosis, pre-location and pin-pointing as suited to network and staff requirements. For the industrial network referred to above, the most important piece of equipment is the surge generator because, even if a pre-location is not possible for any reason, short routes of several hundred metres can be traversed with the acoustic detector in a relatively short space of time. Very much related to this, and an absolute necessity, is a set of audio frequency tracing and identifying equipment, the receiver of which can accept the ground microphone for acoustic detection.

With the greatest respect to all concerned, it is probably true to say that accurate and up to date records are rarely available on industrial sites. Staff changes and alterations and additions to roads and buildings over the years do not help greatly. Equipment for tracing and identifying is therefore most useful.

Maintenance engineers on even the largest industrial sites are not faced with fault location very often and get little practice at it. Therefore all the equipment purchased should be straightforward to operate. The surge generator comes within this category. Insulation resistance testers and multimeters are in daily use and present no problem. For pre-location, refined PE sets/TDRs and arc reflection equipment are too complicated and totally unnecessary. The situation requires a basic, low cost PE set/ TDR, probably backed up by a resistance bridge or a handy modern contact fault locator for pre-location. A burn-down set may also be favoured but its cost is similar to that of the surge generator. The cost of having both is prohibitive. A surge generator with a low current initial burn capacity would be a sensible choice. For industrial sites therefore a suitable minimum kit would comprise the equipment given in Table 16.1.

The best minimum kit for a supply authority 'district' is somewhat different because the network and staff requirements are themselves very different. (In this context, district means the office or depot in a large urban or mixed urban and rural area from which all routine operation and maintenance is carried out.)

Fault incidence is high. There are low voltage faults daily and MV/HV faults several times per month. High and low voltage cables require different approaches. Because of the high fault incidence, the staff concerned are trained and in practice.

The minimum kit should therefore include basic diagnostic instruments as before plus the full audio frequency tracing/identification and acoustic set with all available accessories, together with a surge generator, this time

Table 16.1 Minimum kit needed for industrial sites

Function	Equipment	Approximate cost (% of maximum outlay)
Diagnosis	IR tester and multimeter	0.2
Pre-location	Contact fault locator	1.0
	Basic PE/TDR	1.5
Pin-pointing	Surge generator	9.0
Trace/identification/ acoustic	Audio frequency set plus accessories	5.0
	Total	16.7
	Optional	
Treatment	Burn-down set	7.0

with the transient capture interface and monitor for arc reflection, PE/TDR and impulse current pre-location. Burn-down sets would always have been included in the past but their use has declined with the adoption of impulse current and arc reflection methods. They can be considered as an option.

This combination is best mounted in a small van, with or without purpose-built cable drums. The remaining portable equipment is carried in cupboards in the same vehicle. To this should be added a mid range portable PE set/TDR, a contact fault locator, a fault re-energizing device (FRED) and, if required, a sheath fault location set. It is useful if the PE set/TDR has a trigger facility for use on live low voltage networks (see Chapter 9, Fig.9.7 and the associated text).

The list, then, for a supply authority 'district' resembles more a complete kit than a minimum kit. However, the actual minimum kit required depends very much on how the various central, area and district teams are equipped and organized. Often there are historical factors to be taken into consideration as well. Therefore, in the suggested supply authority minimum kit given in Table 16.2, any of the items marked * can be taken out of the list depending on these organizational aspects.

The minimum kit for a company or authority covering road and street lighting only is dictated by different priorities again. The troubleshooting teams probably carry out jointing as well as fault location and the approach defaults too readily to cutting and testing if no effective instruments are provided. On the other side of this coin is the fact that complex and expensive instruments are not necessary; nor can their issue to all teams be justified.

Table 16.2 Minimum kit suggested for a supply authority

Function	Equipment	Approximate cost (% of maximum outlay)
Diagnosis	IR tester and multimeter	0.2
Pre-location (portable)	Contact fault locator*	1.0
	PE/TDR*	3.0
Treatment	Burn-down set*	7.0
Pin-pointing (including pre-location)	Surge generator interface and monitor	25.0
Trace/identification/ acoustic	Audio frequency set plus accessories	5.0
Re-energize	FRED*	2.5
Earth contact location	Pool of potential equipment*	4.0
	Total	47.7
	Less items marked * at 17.5	30.2

Table 16.3 Minimum kit for road and street lighting teams

Function	Equipment	Approximate cost (% of maximum outlay)
Diagnosis	IR tester and multimeter	0.2
Pre-location	Contact fault locator	1.0
	Basic PE/TDR	1.5
Trace/identification/ acoustic	Audio frequency set plus accessories	5.0
	Total	7.7
	Optional	
Pin-pointing	Small surge generator	5.0

If distances are relatively short and access is good, the best compromise is to use small handy cost effective PE sets/TDRs and contact fault locators for pre-location and forego pin-pointing by surge generator. Another reason for not employing large surge generators is *safety*. Higher voltages are available than should be used on the system and the operator can be tempted to use them. A useful solution is to purchase a small surge generator with lower voltages and keep this for 'on demand' use by well-trained staff for awkward faults.

Audio frequency gear is necessary for tracing and occasional pin-pointing.

A minimum kit for road and street lighting teams is given in Table 16.3.

The minimum kit for communications fault location must be considered in two distinct categories: equipment for large telecommunications companies and authorities and equipment for all other bodies having communications, control and information networks, such as:

- Electricity
- Gas
- Water
- Oil and petrochemical
- Airport
- Harbour
- Industrial
- etc.

As indicated earlier, the former tend to use faultmen/jointers or teams troubleshooting on a daily basis, equipped with basic diagnostic, pre-location and tracing equipment. Back-up staff with more refined instruments are called in to tackle the more awkward situations. The minimum kits for these teams could be those given in Table 16.4, where the low costs are due to bulk purchase of standard issue equipment, and Table 16.5.

Table 16.4 A minimum kit for a faultsman/jointer team

Function	Equipment	Approximate cost (% of maximum outlay)
Diagnosis	Multifunction meter	0.15
Pre-location	Contact fault locator	0.2
	Basic PE/TDR	0.5
Trace/identification	Issue tone tester	0.15
	Total	1.0

Fault location and maintenance engineers in the latter group have to deal with communications fault location on telecommunications, telemetry, information, data, control and composite cables and, because they are on the 'technical' side, may well be called upon to deal with power cable faults as well. Their minimum kit therefore ought to be a combination of the kits for both the above telecommunications teams with options on small burn-down sets which can also be used as the source for pool of potential ground contact fault location, and small surge generators for pin-pointing. The minimum kit becomes as given in Table 16.6.

16.2 Portable equipment versus test trailers and test vans

There are many factors involved in deciding whether portable equipment should be purchased, whether a tailor-made test van should be commissioned or whether a trailer should be constructed as a compromise. Among them are:

- Cost
- Organization of fault location groups and teams

Table 16.5 A minimum kit for a back-up team

Function	Equipment	Approximate cost (% of maximum outlay)
Diagnosis	IR tester, multimeter	0.2
Pre-location	Resistance bridge	3.0
	Comprehensive PE/TDR	13.0
Treatment	Small burn-down set/ fault enhancer	4.0
Trace/identification/ pin-point	Comprehensive audio frequency set	4.0
	Total	24.2

Table 16.6 A minimum kit for authorities with telecommunications networks

Function	Equipment	Approximate cost (% of maximum outlay)
Diagnosis	IR tester, multimeter	0.2
Pre-location	Resistance bridge	3.0
	Comprehensive PE/TDR	13.0
Treatment	Small burn-down set/ fault enhancer	4.0
Trace/identification/ pin-point	Comprehensive audio frequency set	4.0
Pin-pointing	Small surge generator	5.0
	Total	29.2

- Distances involved
- Terrain
- Climate
- HV testing requirements
- Power requirements
- Operational safety

A fully equipped test van may be required where:

1 It is vital to keep outage times short, for instance on networks with many radial feeders on which no back feeds can be made.
2 There is a high incidence of faults in a relatively small area.
3 In a wider area, there is demand for expert assistance on difficult fault locations. This type of duty is usually found when a 'sun and planets' system is in place. This is a very efficient way of handling fault location in a large electricity distribution company or authority with a central headquarters and outposted districts or depots. The latter are equipped with portable equipment costing approximately 20 per cent and capable of finding most faults on the LV and MV networks. The centrally based tailor-made test van is then called for when depot staff encounter a particularly difficult fault, the location of which is beyond their capability or the capability of their equipment.
4 Specialized attention is required for high voltage fault location in a difficult environment where an air conditioned or a heated mobile laboratory with built-in generator is essential for efficient working.
5 Where the workload for high voltage testing coincides with the workload or organizational requirements for fault location.

Test trailers are specified when:

1 Expenditure has to be kept to a minimum.
2 It is not possible to commit a large expensive vehicle permanently to fault location.
3 There is a fleet of large vehicles all fitted with towing hitches normally towing other trailers, e.g. for jointing.
4 There is no requirement for plan tables, cupboards, heating/air conditioning, locked-off HV test area or motor-driven generator.

16.3 Test van specifications – general

There is a conflict of interests between the efficient use of fault location equipment and the efficient use of a vehicle. The vehicle covers a much lower mileage on fault location work than it would if used for other purposes. Also, the replacement cycles of test equipment and vehicle may or may not coincide.

In the past tailor-made vehicles have been just that, tailor-made, where the test instruments are 'built in' as opposed to being 'fitted in' and are an integral part of the vehicle. Thus the vehicle was operated in some cases for more than 10 years. In these days of relatively small technically refined equipment there is a lot to be said for a *modular* approach to fitting out test vans whereby the *specification* rather than the *build* is 'tailor-made'.

The only aspects of the build that are specific to the vehicle are:

- Insulation
- Fitted tables and cupboards
- Dividing walls
- Air conditioning (if fitted)
- Generator and power take-off (if fitted)
- Special doors, seating and cabin arrangements

When respecifying there is therefore more choice in considering whether to use the same vehicle or another one because the first vehicle can be modified either for the fitting of fault location gear or for another purpose altogether.

The ultimate 'modular' approach is the building of a complete fully equipped fault location cabin in a conventional *container*. This is a neat solution when a given company/authority has several platform vehicles in use for carrying containers for other purposes. There is then absolutely no conflict between the efficient use of the vehicle or the test equipment.

It is not intended here to give actual specifications, but simply to highlight items and factors that the specifying engineer needs to take into account when making a test vehicle specification. A basic consideration is *voltage*, that is to say, whether or not the vehicle is to be used for HV *testing* as well as routine fault location. If so, this will entail the provision of a lockable HV test compartment having a single-pole HV d.c. source of over 100 kV as well as the single-pole or three-pole outputs at around 70/80 kV for fault location purposes alone. If a high voltage single-pole output is incorporated, it must have its own discharge resistor and discharge spark gap. This in turn creates a space requirement due to the physical size of the various components and their safety clearances. This being the case, it is usually worth while incorporating an HV capacitor bank and adjustable spark gap as a surge generator for pin-pointing on HV cable faults with a high voltage withstand.

Another major consideration is *switching*; i.e. whether the MV output from the surge generator and burn-down units as well as LV items such as the PE set/TDR, audio frequency tracing generator and diagnostic instruments, can be routed to one or more poles which are connected to the three cores of a cable. Alternatively, the LV diagnostic and test instruments can be routed via three poles and the MV test connections limited to one pole.

The selection of one particular MV instrument such as a surge generator to a particular output pole is normally made by remote electrical or manual control from the test desk/panel. The switch itself is air, oil or SF6 gas insulated and safely mounted below or behind the control console.

Closely allied to these requirements of voltage and instrument selection is the matter of output cables. Cable drums are mounted at the rear of the vehicle, which carry:

- A screened single-core HV cable with a safe source end plug and socket connection arrangement capable of carrying up to around 130 kV
- A single- or three-core screened MV cable with safe source end connection arrangements and rated at above 50 kV
- An earthing cable of adequate cross-section fitted with connecting ferrules every few metres along its length so that it need not always be fully run out, the length of the earth connection thereby being kept to a minimum
- A smaller auxiliary earth wire for monitoring vehicle-to-earth voltage
- An auxiliary multicore cable for diagnostic, pre-location and tracing connections
- A low voltage supply cable
- A cable for the remote emergency off switch if specified

A normal maximum run-out length for these cables is 50 m and the heavier cable drums are usually motor driven.

Safety systems are of paramount importance and can comprise some or all of the following features:

- Lockable mains on/off switch
- Emergency off switches in the control cabin and at the end of the test cable
- Limit switches which make when the rear doors are closed
- A voltage operated relay which opens if a dangerous voltage appears between ground and the test vehicle body
- A current operated relay which monitors the integrity of the test earth return loop.

It is normal for every operation and instrument selection to be made by remote switch or push button from the main control console. This means the provision of a low voltage d.c. control circuit incorporating suitable relays to effect these selections. Therefore the contacts of all the above-mentioned safety devices are usually simply connected in series in this circuit so that, even if only one of these is open, nothing can be operated. Pilot lamps indicate the state of all the safety devices.

Maximum power requirement is a very critical factor. This determines the size and rating of the low voltage supply cable and the rating of low voltage input switches and protection devices on the control console. The heaviest loads are required by burn-down sets, both for normal burn-down and burn-down associated with arc reflection. The van heating or air conditioning requirement is also high.

The maximum connected load decides the size of any built-in or portable generator which may be specified. Built-in generators of up to 9 kVA can be driven by the vehicle engine from a power take-off. Diesel engines are much easier to control and regulate than petrol engines. It is clear that auxiliary power requirements need to be known before a particular make and type of vehicle is specified. Similarly, air conditioning and insulation requirements should be gone into thoroughly before specifying a vehicle, as these can affect space available inside the vehicle and increase the overall external height. The ground clearance requirements dictated by terrain should also be carefully considered.

Matters of comfort and convenience should not be considered as luxuries. More correct and speedy fault locations will be made if the operators can work in comfort with everything to hand. Furthermore, in many environments around the world such provisions are an absolute necessity. Therefore proper provision should be made for:

- Seating
- Cupboards
- Drawers
- Plan table space

334 UNDERGROUND CABLE FAULT LOCATION

Fig. 16.1 Typical fault location test van (*source:* Seba Dynatronic GmbH, Germany)

- Power points
- Lighting

Figure 16.1 is the rear view of a typical fault location test van as supplied by Seba Dynatronic GmbH.

PART SIX

The future of fault location

In this chapter, matters of fact are faithfully reported, but it must be emphasized that all views and opinions expressed are those of the author.

17

Pressure for solutions, technology and known trends

There has been, is and always will be pressure for manufacturers to produce the perfect fault location instrument, the ultimate 'black box' that will find all types of fault. The faults themselves and network, environmental and organizational circumstances vary so much that it is most unlikely such a device will ever be produced. However, as has been stated already, electronics and computer technology is so advanced that solutions exist now or will soon exist for many of the problems that still bedevil the fault location engineer.

The experienced user has ideas but rarely the knowledge and resources to develop new equipment, while the manufacturer has expertise, knowledge and resources and does not always appreciate some of the factors involved in field testing. Therefore, new fault location techniques could be developed if formal cooperation and joint ventures were entered into by large power and telecommunications organizations and manufacturers and universities. There has always been some activity in these areas but rarely on a proper commercial basis.

Pressure for developing new methods can well derive from the will to improve techniques *per se*. Curiously, though, the main drive is likely to come from purely commercial considerations. There is an accelerating trend to reduce costs, very often by using cleverer instruments to 'deskill' a technical operation such as fault location.

The four main areas requiring investigation are perhaps:

- The analysis and interpretation of pre-location data acquired for display in the PE/TDR, impulse current and arc reflection techniques
- Partial discharge detection and location
- Pin-pointing techniques
- Low voltage methods

Taking the first area, data acquisition, even from high voltage transients, is well advanced and display techniques and quality are excellent. However, the analysis and interpretation is presently carried out by the human being looking at a monitor screen. This is exactly the area in which computer techniques can be used for data analysis and, indeed, several people are currently working on this approach. Though this will take a lot of the fun out of fault location, nothing in life does or should stand still and a quantum leap in this direction is likely.

Research has been carried out by Professor K. Warwick of the Department of Cybernetics, University of Reading, and K.K.Kuan of the Department of Engineering, University of Warwick, which is set out in a paper entitled 'Real-time expert system for fault location on high voltage underground distribution cables', presented in the UK *IEE Proceedings Part, C*, Vol.139, No.3, May 1992.

This work uses the transient data produced by impulse current tests and is based on the premise that the interpretation of impulse current traces is heavily dependent on the operators' knowledge and experience. The solution proposed is a hybrid approach in software, wherein resides a structured part capable of performing all the necessary analyses, comparisons and calculations and an unstructured part – the 'expert system'. The operator carries out the normal diagnostic tests and enters these into the computer which then advises what test is to be carried out on what phase. When the resulting transient waveform is captured, the data are entered into the expert system database and analysed in conjunction with the diagnostic information and the type and location of the fault is read out. A key feature of this approach is that the expert system can be progressively encoded with data from non-typical faults and may, in future, be made capable of self-learning.

Work has also been done by Hathaway Instruments Limited, and an intelligent fault location instrument has been produced which is capable of analysing fault location data from any fault locator with a digital interface, thus reducing the need for specialist operators. This is the Telefault 2000 shown in Fig.17.1.

The instrument is menu driven and has three modes of operation: manual, computer aided and automatic. Thus the operator can choose to carry out a personal interpretation or opt for assistance from the instrument at two other levels. A disc drive is incorporated for saving and loading records, protocols and applications.

With regard to partial discharge (PD), the location of a partial discharge site surely represents the ultimate in cable fault location – the finding of a fault before it occurs! However, as everyone concerned knows, the whole subject of partial discharge is extremely complex.

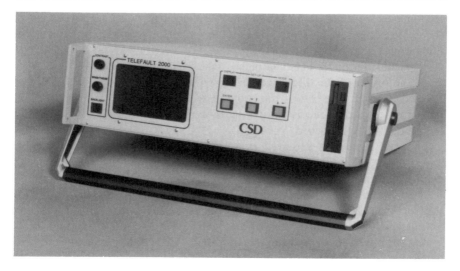

Fig. 17.1 The Telefault 2000 (*source:* Hathaway Instruments Ltd, Hoddesdon, Hertfordshire, UK)

It is possible to detect the presence of discharge and to locate the site by special PE/TDR techniques in a screened environment, and, in certain circumstances, in the field. The whole business, though, is bristling with difficulties. PD is an a.c. phenomenon. Large power transformers are required to excite a sample at 50 Hz. Alternatively, very low frequency (VLF) or special resonance sets can be used. And even when successful measurements have been made, how is it possible to say *when* a certain level of discharge activity will produce a flashover?

In the future a sound approach could be to install on-line monitoring equipment on MV systems at primary substations. The onset of a certain level of activity could then be signalled and the faulty feeder identified. Even if a pre-location can be made, is it then worth digging the cable up on a pre-location in the absence of a pin-pointing method?

The foregoing is just an indication of the very many difficulties that exist but, nonetheless, the fact remains that PD is a forewarning of failure and research must continue into how it can be used to find faults. In fact, a considerable amount of successful work has been done by EA Technology, Capenhurst, Chester, over the last five years in actually 'mapping' the sites of partial discharge in 11 and 33 kV cables in the field.

A VLF discharge free voltage source is employed which is rated at 35 kV and 20 mA. The positive and negative currents are controlled by a waveform generator and feedback control via optical links. Full working voltage at frequencies from d.c. to 10 Hz can be supplied, depending on the length of the cable under test, and the power requirements are within the capacity of any domestic supply point.

A partial discharge event launches fast edged pulses in both directions from the origin of the discharge, as shown in Fig.17.2, which is a record of the waveform at one end of the cable. It will be seen that the time-of-flight measurement technique is similar to that used in other methods of fault location already described.

The raw data thus acquired is used to construct a 'map' of the cable showing the magnitude and position of all the discharge activity on it. An example of this is shown in Fig.17.3.

The system is being used to build up experience to provide a database for the service performance of a variety of paper-insulated cable types and voltages, with the eventual aim of being able to predict the remaining life of cables from tests made and statistical evidence gathered. Significant progress is being made with this at the time of writing.

There is also surely a great need for research and development in pin-pointing techniques. The ultimate 'black box' in this field is, of course, the instrument, which can be walked over the route and bleeps when over the fault. This may be impossible for all kinds of fault but certain pin-pointing methods, such as the pool of potential and acoustic methods, already incorporate aspects of this.

In power engineering, existing pin-pointing methods have been reasonably effective for many years and it is probably true to say that, because of this, research and development in this field has been grossly neglected. Yet a 'yes/no', 'go back/go forward' overground sectionalizing and pin-pointing device could almost make pre-location redundant on short cables. It is always preferable to pin-point if at all possible, as most time and money is wasted digging holes in the wrong place and cutting cables. What about the beleaguered telecommunications engineer, for whom pin-pointing just does not exist in most direct buried fault location situations? There is a definite requirement for a 'back to basics' look at what modern technology can accomplish in detecting overground the *manifestations* of a fault, as listed in Sec. 4.1. Never is this statement more true than in the case of the overground location of faults on complex, multiteed low voltage systems, where pin-pointing is, to say the least, difficult!

It is also absolutely essential that research be carried out on an *actual live network*. The high fault levels found on real systems are necessary to examine real fault situations. Little is to be gained by examining faults on simulated systems. Again this starkly emphasizes the necessity for serious commercial cooperation between instrument manufacturers and electricity supply companies.

PRESSURE FOR SOLUTIONS, TECHNOLOGY AND KNOWN TRENDS

Aside from the necessity for research into pin-pointing techniques on complex low voltage networks, mentioned above, the most promising area for development is perhaps in the use of low voltage transient techniques,

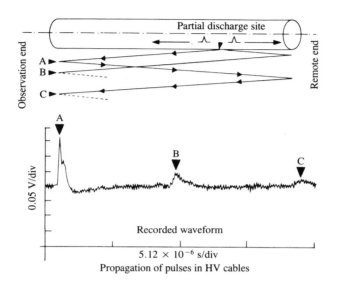

Fig. 17.2 Propagation of partial discharge pulses

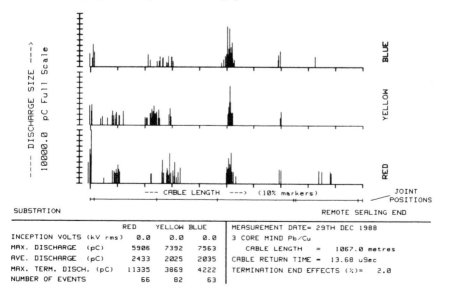

Fig. 17.3 'Map' of partial discharge activity

described in Sec. 9.1, 'transient methods', in conjunction with fault re-energizing devices (FREDs). Several electricity supply companies are using automatic reclosing versions of FRED together with a transient fault locator and achieving good results. This activity is only taking place, however, in certain pockets where keen, interested and knowledgeable engineers have made considerable efforts to address low voltage fault location problems.

There is a perceived need for the few existing transient fault locators to be replaced by cleverer, smaller, cheaper, more user friendly and more effective units in order to transform what is the art of the enthusiast into a routinely successful method for the many. The technology exists.

18

Conclusion

Whether you have read this volume from cover to cover, just browsed or looked up something specific, I do sincerely hope that the book has been and will continue to be useful in enabling you to learn more about cable fault location or in helping you to solve the problems encountered daily in this most difficult branch of electrical engineering.

However, no matter how much knowledge and experience is gained, one thing remains absolutely certain: *cable fault location is not easy*! The more you learn about it, the more humble you become. You may encounter a fault tomorrow and locate it inside one hour – or in a few days. There is no room for complacency or presumption.

The fault will cut you down to size!

However, with logic, knowledge and a little luck, you will prevail.

Further reading

BICC. (1957), *BICC Electrical Conductors*, British Insulated Callendar's Cables Limited, London.

Gale, P.F. (1974), *A New Fault Location Technique for Power Cables*, Bradford University Press, UK.

Gönen, T. (1987), *Electrical Power Distribution System Engineering*, McGraw-Hill, Singapore.

Jäckle, E. (1966), *Elektronische Rohr-und Kablesuche* (Bestellnummer 3390K), Franckh'sche Verlagshandlung, W. Keller and Company, Stuttgart, Germany.

Kurtz, E.B. and Shoemaker, T.M. (1986), *The Lineman's and Cableman's Handbook*, McGraw-Hill, New York.

McAllister, D. (Ed).(1982), *Electric Cables Handbook*, Granada, London.

Index

A frame, 140–144, 190–191, 219
Acoustic, 113–125, 287
Acoustic signal, 124
Aerial coil, 174–175
Airfield lighting, 282–287
Amplifier, 15, 121
Arc, 63, 216, 227, 229, 230, 265
Arc drop/voltage, 7, 216
Arc reflection method, 97–109
Attenuation, 44, 53, 288–293
Audio frequency, 15, 139–151, 169–191, 194–196, 265

Backfilling, 152, 154
Banger, 115, 228
Ballistic galvanometer, 285–286
Bartec, 261
'Before and after' method, 22, 220–221
Bimec, 144
Bolometer, 277
Bridge (*see* Resistance bridge), 23–42, 257–262
Budget, 321–324
Burn down, 9, 60–68, 251

Cable:
  combined neutral and earth (CNE), 300
  consac, 301
  cross-linked polyethylene (XLPE), 12, 298–304
  ethylene propylene rubber (EPR), 302
  extra high voltage (EHV), 230–234
  gas filled, 230, 235
  heating, 273–279
  high molecular density thermoplastic polyethylene (HMWPE), 301–304
  high voltage (HV), 230–233
  lighting, 280–287
  medium voltage (MV), 230–233
  oil filled, 4, 230, 235
  polyethylene (PE), 298–300, 302
  polyvinyl chloride (PVC), 297, 302
  pressurized, 248
  urban residential distribution (URD), 209, 211, 301, 302, 304
  waveconal, 302
Cable gun spike, 192–194
Cable tiles, 154
CATV, 288
Caution notice, 202–203
Central office (*see* Exchange), 246
Characteristic impedance, 43–47, 51–52
Circuit breaker, 203
Cladding, 288–292
Cocktail party effect, 121
Cold tail, 274
Comparison of traces, 53–54
Confirmation, 16, 152
Connector, 291, 293
Continuity, 20–21, 84
Control systems, 267
Coupling, 172
Cross-sectional area/gauge, 30–32, 261
Cut and test, 160, 162, 213–214

Danger notice, 202–203
Data acquisition, 338
DC impulse method, 196–197
Depth checking, 176–177, 181–182
Diagnosis, 10–12, 17–22
Digital fault locator, 71–73
Double coil assembly, 149–150
Ducts, 124, 154, 187, 247

Electromagnetic pin-pointing, 125–128
Equivalent length, 30–31
Equivalent circuit (of fault), 75
Excavation, 152–154
Exchange, 246

Fault (type):
  battery, 42, 250, 257
  break (open circuit), 4, 19–21, 43
  contact, 3, 23
  core-to-core, 3, 145

347

core-to-sheath, 3, 148
crimping, 4–5, 253–254
cross-talk, 5, 55, 254
earth, 3
flashing, 5, 68
foreign battery, 42, 250, 257
ground contact, 3, 128–144
ingress of moisture, 4–5, 67, 251–253
open circuit, 4, 19–21, 43
partial break, 4
series, 4, 44–48
serving, 3
sheath, 3, 236–237
shunt, 3, 23
transitory, 6–7
Fault enhancer, 252
Fault location management, 223–224
Fault re-energizing device (FRED), 215–216, 223–226, 342
Field turbidity method, 150–151
Flow charts, 162–165
Flux linkage, 169–170
Four-wire test, 29
Fresnel reflection, 292
Fully filled cable, 249

Galvanic connection, 172–174
Graded index fibre, 289
Gradient test, 216
Ground survey, 176, 185

Halving, 258
Hazardous areas, 288
Hilborn loop test, 33, 160–162

Identification, 194–198
Impulse current method, 73–96
Inductive coupling, 174–176
Information systems, 267–269
Insulation displacement connector (IDC), 258
Insulation resistance (IR), 17–18
Inverted loop test, 34, 37, 42
Ionization delay, 70–90
Isolator, 203

Joint location, 176, 190–191
Joules, 114
Jelly filled, 249

Kick test, 247
Kirchhoff's current law, 178

Laser, 288, 292
Lattice diagram, 78–81
LED, 288
Linear coupler, 73–74
Local area network (LAN), 267–268
Local distribution network, 248
Local junction network, 248
Local loop, 249
Log sheet, 12–14
Loop-on, loop-off, 91–96, 233–235
Loss budget, 291
Losses, 290–291
Lotec, 216, 221

Magnetic field, 136, 169–171
Magnetic signal, 121, 124
Magnetometer, 136–139
Marker tape, 154
Metal detection, 176, 188–189
Metallic, 169, 288, 292
Micro bend, 292
Microphone, 15, 120–124
Minimum kit, 325–330
Mode, 288–289
Multimode, 288–289
Multiple earth system, 210
Murray loop test, 23, 34, 36, 42

Optical fibre, 246, 288–293
Optical time domain reflectometer (OTDR), 291–293

Partial discharge, 7–8, 338–341
Permit to work, 203–204
Picocoulomb, 7
Pin-pointing, 15, 112–151
Polaroid, 71, 155
Pool of potential, 128–136, 140
Pre-location, 12, 23–105
Profile alignment system, 291
Pulse echo (PE), 42–60
Pulse code modulation (PCM), 248
Pulse width 53

Rayleigh scattering, 291–292
Receiver, 177
Recording, 16, 155
Reflection factor, 44–47
Refractive index, 288–289
Reinstatement, 152, 154
Relative permittivity, 43
Relaxation test, 81–83, 85–86, 231

Repairs, 156
Reporting, 16, 157
Resistance bridge (*see* bridge), 23–42, 257–262
Resonance set, 339
Re-split, 51, 255–256, 266
Ring, 256

Safety, 10, 201
Sanction for test, 202–203, 205
Screening, 297–304
Shock wave discharge, 15, 112–124
Single mode fibre, 290
Splice, 291–293
Split, 51, 254–256, 266
Splitter box, 94
Spoil, 153
Step index fibre, 288–289
'Sun and planets' system, 330
Surge energy, 114–115
Surge generator, 113–114
Surge wave, 114

Tee branches, 32–33, 51–52, 147, 231–232, 283
Telecommunications, 245–266
Test trailer, 329–331
Test van, 329–334

Thermal imager (gun), 277–278
Third party, 16, 155
Thumper, 115
Time domain reflectometer (TDR), 42–60
Time lapse, 77–78
Tip, 256
Transgradient method, 216
Transient methods, 68–105, 221–223
Transmitter (generator) 172–176, 182
Transmitter tongs, 174–176, 184
Treeing, 7–8
Trefoil, 230–231, 298
Trunk network, 248
Turbidity, 150–151, 218
Twist method, 68, 148–149, 190, 194, 217

Unearthed systems, 233, 269
URD cables (*see* Cables)

Velocity factor, 43
Very low frequency (VLF), 339
Velocity of propagation, 42–43, 48–49
Voltage discriminator cables (VODCA), 199–200

Water tree, 302–303